Ben Stacy Jerrik (Ed.)

Caproni Ca.122

Ben Stacy Jerrik (Ed.)

Caproni Ca.122

Bomber, Prototype, Monoplane, Undercarriage

Part Press

Imprint

Permission is granted to copy, distribute and/or modify this document under the terms of the GNU Free Documentation License, Version 1.2 or any later version published by the Free Software Foundation; with no Invariant Sections, with the Front-Cover Texts, and with the Back- Cover Texts. A copy of the license is included in the section entitled "GNU Free Documentation License".

All parts of this book are extracted from Wikipedia, the free encyclopedia (www.wikipedia.org).

You can get detailed informations about the authors of this collection of articles at the end of this book. The editors (Ed.) of this book are no authors. They have not modified or extended the original texts.

Pictures published in this book can be under different licences than the GNU Free Documentation License. You can get detailed informations about the authors and licences of pictures at the end of this book.

The content of this book was generated collaboratively by volunteers. Please be advised that nothing found here has necessarily been reviewed by people with the expertise required to provide you with complete, accurate or reliable information. Some information in this book maybe misleading or wrong. The Publisher does not guarantee the validity of the information found here. If you need specific advice (f.e. in fields of medical, legal, financial, or risk management questions) please contact a professional who is licensed or knowledgeable in that area.

Any brand names and product names mentioned in this book are subject to trademark, brand or patent protection and are trademarks or registered trademarks of their respective holders. The use of brand names, product names, common names, trade names, product descriptions etc. even without a particular marking in this works is in no way to be construed to mean that such names may be regarded as unrestricted in respect of trademark and brand protection legislation and could thus be used by anyone.

Cover image: www.ingimage.com
Concerning the licence of the cover image please contact ingimage.

Publisher:
Part Press is a trademark of
International Book Market Service Ltd., 17 Rue Meldrum, Beau Bassin, 1713-01 Mauritius
Email: info@bookmarketservice.com
Website: www.bookmarketservice.com

Published in 2012

Printed in: U.S.A., U.K., Germany. This book was not produced in Mauritius.

ISBN: 978-613-6-13991-3

Contents

Caproni_Ca.122

Ca.122	
Role	Bomber and military transport
Manufacturer	Caproni
Status	Prototypes only

The **Caproni Ca.122** was a prototype bomber and military transport aircraft built in Italy in the mid-1930s. It was a conventional low-wing monoplane with fixed undercarriage. Power was provided by twin Gnome-Rhône 14K radial engines. No production ensued for either this or the a 28-seat airliner version designated **Ca.123**.

Specifications (Ca.123)

General characteristics

- **Crew:** 3: Pilot, co-pilot, and radio operator
- **Capacity:** 28 passengers
- **Length:** 18.19 m (59 ft 8 in)
- **Wingspan:** 27.87 m (91 ft 5 in)
- **Height:** 6.00 m (19 ft 8 in)
- **Wing area:** 91.6 m^2 (986 ft^2)
- **Empty weight:** 5,300 kg (11,686 lb)
- **Gross weight:** 8,800 kg (19,404 lb)
- **Powerplant:** 2 × Gnome-Rhône 14K, 650 kW (870 hp) each

Performance

- **Maximum speed:** 340 km/h (211 mph)
- **Range:** 1,500 km (932 miles)
- **Service ceiling:** 7,530 m (27,400 ft)
- **Rate of climb:** 0.1 m/s (13 ft/min)

References

- Taylor, Michael J. H. (1989). *Jane's Encyclopedia of Aviation*. London: Studio Editions. pp. 234.
- Thompson, Jack (1963). *Italian civil and military aircraft 1930-1945*. Fallbrook: Aero Publishers. pp. 86.

Bomber

A **bomber** is a military aircraft designed to attack ground and sea targets, by dropping bombs on them, or – in recent years – by launching cruise missiles at them.

The B-17 Flying Fortress is a heavy bomber from World War II.

Classification

Strategic

Strategic bombing are heavy bombers primarily designed for long-range bombing missions against strategic targets such as supply bases, bridges, factories, shipyards, and cities themselves, in order to damage an enemy's war effort. Current examples include the strategic nuclear-armed strategic bombers: B-2 Spirit, B-52 Stratofortress, Tupolev Tu-95 'Bear', Tupolev Tu-22M 'Backfire'; historically notable examples are the: Gotha G, Avro Lancaster, Heinkel He-111, Junkers Ju 88, B-17 Flying Fortress, B-24 Liberator, B-29 Superfortress, and Tupolev Tu-16 'Badger'.

A Tupolev Tu-160 strategic bomber.

Tactical

Tactical bombing , aimed at enemy military units and installations, is typically assigned to smaller aircraft operating at shorter ranges, typically along the troops on the ground or sea. This role is filled by various aircraft tactical bomber classes, as different as light bombers, medium bombers, dive bombers, interdictors, fighter-bombers, ground-attack aircraft, multirole combat aircraft, among others. Current examples: F-15E Strike Eagle, F/A-18 Hornet, Sukhoi Su-34 'Fullback', Chengdu J-10, Xian JH-7, Dassault-Breguet Mirage 2000, and the Panavia Tornado; historical examples: Ilyushin Il-2 *Shturmovik*, Heinkel He 111, Dornier Do 17, Dornier Do 215, Junkers Ju 88, Junkers Ju 87 *Stuka*, P-47 Thunderbolt, Hawker Typhoon, and F-4 Phantom II.

History

1911–1939

The first use of an air-dropped bomb was carried out by the Italians, initially by Lieutenant Giulio Gavotti,[1] in their 1911 war for Libya. In 1912 Bulgarian Air Force pilot Christo Toprakchiev suggested the use of airplanes to drop "bombs" (as grenades were called in the Bulgarian army at this time) on Turkish positions. Captain Simeon Petrov developed the idea and created several prototypes by adapting different types of grenades and increasing their payload.[2] On October 16, 1912, observer Prodan Tarakchiev dropped two of those bombs on the Turkish railway station of Karaagac (near the besieged Edirne) from an Albatros F.II airplane piloted by Radul Milkov.

Caproni Ca.32, Italian heavy bomber of World War I

After a number of tests Petrov created the final design, with improved aerodynamics, an X-shaped tail and impact detonator. This version was widely used by the Bulgarian Air Force during the siege of Edirne. Later a copy of the plans was sold to Germany and the bomb, codenamed "Chathaldza" ("Чаталджа", after the strategic Turkish town of Çatalca) remained in mass production until the end of World War I.

The weight of the bomb was 6 kilograms (**unknown operator: u'strong'** lb); on impact it created a crater 4–5 metres (**unknown operator: u'strong'unknown operator: u'strong'unknown operator: u'strong' unknown operator: u'strong'**) wide and about 1 metre () deep.

The first two-engined bomber, the British Handley Page Type O

E. Mark cites Caproni Ca 30 and Bristol TB.8, both of 1913, as one of the first of heavier-than-air aircraft purposely designed for bombing.[3]

During World War I, the Germans used Zeppelins as bombers since they had the range and capacity to carry a useful bomb load from Germany to England. With advances in aircraft design and equipment, they were joined by larger multi-engined biplane aircraft on both sides for long range strategic bombing especially by night. The majority of bombing was still done by one-engined biplanes with one or two crew-members flying short distances to attack the enemy lines and immediate hinterland.

The world's first four-engined bomber was the Russian Il'ya Muromets created in 1914 and successfully used in World War I.

World War II

During World War II bombers often looked dramatically different from other aircraft. Because of the lack of power in aircraft engines at the time, bombers needed to have multiple engines in order to carry a reasonable load, in turn leading to much larger aircraft.

United States Navy SBD Dauntless dive bomber

A Grumman TBF Avenger torpedo bomber

With engine power as a major limitation, combined with the desire for accuracy and other operational factors, bomber designs tended to be tailored to one particular role. By the start of the war this included

- dive bomber
- light bomber, medium bomber and heavy bomber
- torpedo bomber
- interdictors
- specialized ground attack aircraft designs

Bombers are not intended to actively engage in combat with other aircraft. The majority have been relatively large and unmaneuverable – although some smaller designs have been used as the basis for specialist fighters, such as night fighters.

Cold War

At the start of the Cold War, bombers were the only means to take nuclear weapons to enemy targets, and had the role of deterrence. With the advent of guided air to air missiles, bombers needed to avoid interception. High speed and high altitude flying became a means of evading detection and attack. Designs such as the English Electric Canberra could fly faster or higher than contemporary fighters. When surface to air missiles became capable of hitting high flying aircraft bombers, bombers used flight at low altitude to evade radar detection.

A B-58 Hustler Mach 2 capable supersonic bomber

Once "stand off" nuclear weapon designs were developed, bombers did not need to pass over the target at high altitude to make an attack; they could fire and turn away to escape the blast. Nuclear strike aircraft were generally finished in bare metal or anti-flash white to avoid any flash damage.

The need to drop conventional bombs remained in conflicts with a non-nuclear powers, such as the Vietnam War or Malayan Emergency.

The development of large strategic bombers stagnated in the later part of the Cold War because of spiraling costs and the development of the Intercontinental ballistic missile (ICBM) – which was felt to have equal deterrent value while being   much   more   difficult   to   intercept.

Because of this, the United States Air Force XB-70 Valkyrie program was cancelled in the early 1960s; the later B-1B Lancer and B-2 Spirit aircraft entered service only after protracted political and development problems. Their high cost meant that few were built and the 1950s-designed B-52s continued in use into the 21st century. Similarly, the Soviet Union used the intermediate-range Tu-22M 'Backfire'in the 1970s, but their Mach 3 bomber project came to naught. The Mach 2 Tu-160 'Blackjack' was built only in tiny numbers, leaving the 1950s Tupolev Tu-16 and Tu-95 'Bear' heavy bombers to continue being used into the 21st century.

US Navy F-14 Tomcat escorts Tu-95 Bear D during 1985 NATO exercise Ocean Safari

The Avro Vulcan was part of the RAF V bomber force

The British strategic bombing force largely came to an end when the V bomber force was phased out; the last of which left service in 1983. The French Mirage IV bomber version was retired in 1996, although the Mirage 2000N and the Rafale have a taken on this role. The only other nation that fields strategic bombing forces is the People's Republic of China, which has a number of Xian H-6s.

Modern era

In modern air forces, the distinction between bombers, fighter-bombers, and attack aircraft has become blurred. Many attack aircraft, even ones that *look* like fighters, are optimized to drop bombs, with very little ability to engage in aerial combat. Indeed, the design qualities that make an effective low-level attack aircraft make for a distinctly inferior air superiority fighter, and vice versa. Conversely, many fighter aircraft, such as the F-16, are often used as 'bomb trucks,' despite being designed for aerial combat. Perhaps the one meaningful distinction at present is the question of range: a bomber is generally a long-range aircraft capable of striking targets deep within enemy territory, whereas fighter bombers and attack aircraft are limited to 'theater' missions in and around the immediate area of battlefield

Transferring a 2,000 pound JDAM to a lift truck for loading onto a B-1B Lancer supersonic strategic bomber in Southwest Asia in 2007.

combat. Even that distinction is muddied by the availability of aerial refueling, which greatly increases the potential radius of combat operations.

Plans in the U.S. and Russia for successors to the current strategic bomber force remain only paper projects, and political and funding pressures suggest that they are likely to remain so for the foreseeable future. In the U.S., current plans call for the existing USAF bomber fleet to remain in service until the mid-to-late 2020s, with the first possible replacements becoming operational in 2018.[4] After this bomber the U.S. is also thinking of another bomber in 2037. The 2018 bomber will be made in small quantities as it will be a transition aircraft for this 2037 bomber. The 2018 bomber was, however, required to provide an answer to the fifth generation defense systems (such as SA-21

Growlers, bistatic radar and Active Electronically Scanned Array radar). Also, it was chosen to be able to stand strong against rising superpowers (China, India) and other countries with semi-advanced military capability (Iran). Finally, a third reason was long-term air support for areas with a low threat level (Iraq, Afghanistan). The latter was referred to as close air support for the global war on terror (CAS for GWOT). The 2018 bomber would thus be able to stay for extended periods on a same location (called persistence).[5] Also, the 2018 bomber and later bombers could be automated.

See also

- Carpet bombing
- Strategic bomber
- Aerial bombing of cities
- Aerial interdiction
- Offensive counter air
- Fighter

References

[1] Johnston, Alan (10 May 2011). "Libya 1911: How an Italian pilot began the air war era" (http://www.bbc.co.uk/news/ world-europe-13294524). BBC News. . Retrieved 2011-05-23.

[2] "Bulgarian Air Force History 1912-1913(in Bulgarian)" (http://aviation.zonebg.com/istoria/balcan-war/index.php). . Retrieved 2007-11-25.

[3] Mark (1995-07). *Aerial Interdiction: Air Power and the Land Battle in Three American Wars* (http://books.google.com/ books?id=kr2Gc7btCxEC&pg=PA9). pp. 9–10. ISBN 9780788119668. .

[4] "USAF may seek supersonic and unmanned capabilities for bomber" (http://www.janes.com/news/defence/air/jdw/jdw071018_1_n. shtml). . Retrieved 2007-11-25.

[5] Persistence in 2018 bomber (http://www.flightglobal.com/articles/2007/06/12/214539/ speed-bump-usaf-sets-modest-goals-for-new-bomber.html)

Prototype

A **prototype** is an early sample or model built to test a concept or process or to act as a thing to be replicated or learned from. It is a term used in a variety of contexts, including semantics, design, electronics, and software programming. A prototype is designed to test and trial a new design to enhance precision by system analysts and users. Prototyping serves to provide specifications for a real, working system rather than a theoretical one.[1]

The word *prototype* derives from the Greek πρωτότυπον (*prototypon*), "primitive form", neutral of πρωτότυπος (*prototypos*), "original, primitive", from πρῶτος (*protos*), "first" and τύπος (*typos*), "impression".[2]

Semantics

In semantics, prototypes or *proto instances* combine the most representative attributes of a category. Prototypes are typical instances of a category that serve as benchmarks against which the surrounding, less representative.

Design and modeling

In many fields, there is great uncertainty as to whether a new design will actually do what is desired. New designs often have unexpected problems. A prototype is often used as part of the product design process to allow engineers and designers the ability to explore design alternatives, test theories and confirm performance prior to starting production of a new product. Engineers use their experience to tailor the prototype according to the specific unknowns still present in the intended design. For example, some prototypes are used to confirm and verify consumer interest in a proposed design whereas other prototypes will attempt to verify the performance or suitability of a specific design approach.

In general, an iterative series of prototypes will be designed, constructed and tested as the final design emerges and is prepared for production. With rare exceptions, multiple iterations of prototypes are used to progressively refine the design. A common strategy is to design, test, evaluate and then modify the design based on analysis of the prototype.

In many products it is common to assign the prototype iterations Greek letters. For example, a first iteration prototype may be called an "Alpha" prototype. Often this iteration is not expected to perform as intended and some amount of failures or issues are anticipated. Subsequent prototyping iterations (Beta, Gamma, etc.) will be expected to resolve issues and perform closer to the final production intent.

In many product development organizations, prototyping specialists are employed - individuals with specialized skills and training in general fabrication techniques that can help bridge between theoretical designs and the fabrication of prototypes.

Basic prototype categories

There is no general agreement on what constitutes a "prototype" and the word is often used interchangeably with the word "model" which can cause confusion. In general, "prototypes" fall into five basic categories:

Proof-of-Principle Prototype (Model) (in electronics sometimes built on a breadboard). A Proof of concept prototype is used to test some aspect of the intended design without attempting to exactly simulate the visual appearance, choice of materials or intended manufacturing process. Such prototypes can be used to "prove" out a potential design approach such as range of motion, mechanics, sensors, architecture, etc. These types of models are often used to identify which design options will not work, or where further development and testing is necessary.

Form Study Prototype (Model). This type of prototype will allow designers to explore the basic size, look and feel of a product without simulating the actual function or exact visual appearance of the product. They can help assess ergonomic factors and provide insight into visual aspects of the product's final form. Form Study Prototypes are often hand-carved or machined models from easily sculpted, inexpensive materials (e.g., urethane foam), without representing the intended color, finish, or texture. Due to the materials used, these models are intended for internal decision making and are generally not durable enough or suitable for use by representative users or consumers.

User Experience Prototype (Model). A User Experience Model invites active human interaction and is primarily used to support user focused research. While intentionally not addressing possible aesthetic treatments, this type of model does more accurately represent the overall size, proportions, interfaces, and articulation of a promising concept. This type of model allows early assessment of how a potential user interacts with various elements, motions, and actions of a concept which define the initial use scenario and overall user experience. As these models are fully intended to be used and handled, more robust construction is key. Materials typically include plywood, REN shape, RP processes and CNC machined components. Construction of user experience models is typically driven by preliminary CAID/CAD which may be constructed from scratch or with methods such as industrial CT scanning.

Visual Prototype (Model) will capture the intended design aesthetic and simulate the appearance, color and surface textures of the intended product but will not actually embody the function(s) of the final product. These models will be suitable for use in market research, executive reviews and approval, packaging mock-ups, and photo shoots for sales literature.

Functional Prototype (Model) (also called a working prototype) will, to the greatest extent practical, attempt to simulate the final design, aesthetics, materials and functionality of the intended design. The functional prototype may be reduced in size (scaled down) in order to reduce costs. The construction of a fully working full-scale prototype and the ultimate test of concept, is the engineers' final check for design flaws and allows last-minute improvements to be made before larger production runs are ordered.

Differences between a prototype and a production design

In general, prototypes will differ from the final production variant in three fundamental ways:

Materials. Production materials may require manufacturing processes involving higher capital costs than what is practical for prototyping. Instead, engineers or prototyping specialists will attempt to substitute materials with properties that simulate the intended final material.

Processes. Often expensive and time consuming unique tooling is required to fabricate a custom design. Prototypes will often compromise by using more variable processes, repeatable or controlled methods; substandard, inefficient, or substandard technology sources; or insufficient testing for technology maturity.

Lower fidelity. Final production designs often require extensive effort to capture high volume manufacturing detail. Such detail is generally unwarranted for prototypes as some refinement to the design is to be expected. Often prototypes are built using very limited engineering detail as compared to final production intent, which often uses statistical process controls and rigorous testing.

Characteristics and limitations of prototypes

Engineers and prototyping specialists seek to understand the limitations of prototypes to exactly simulate the characteristics of their intended design. A degree of skill and experience is necessary to effectively use prototyping as a design verification tool.

It is important to realize that by their very definition, prototypes will represent some compromise from the final production design. Due to differences in materials, processes and design fidelity, it is possible that a prototype may fail to perform acceptably whereas the production design may have been sound. A counter-intuitive idea is that prototypes may actually perform acceptably whereas the production design may be flawed since prototyping materials and processes may occasionally outperform their production counterparts.

In general, it can be expected that individual prototype costs will be substantially greater than the final production costs due to inefficiencies in materials and processes. Prototypes are also used to revise the design for the purposes of reducing costs through optimization and refinement.

It is possible to use prototype testing to reduce the risk that a design may not perform acceptably, however prototypes generally cannot eliminate all risk. There are pragmatic and practical limitations to the ability of a prototype to match the intended final performance of the product and some allowances and engineering judgement are often required before moving forward with a production design.

Building the full design is often expensive and can be time-consuming, especially when repeated several times—building the full design, figuring out what the problems are and how to solve them, then building another full design. As an alternative, "rapid-prototyping" or "rapid application development" techniques are used for the initial prototypes, which implement part, but not all, of the complete design. This allows designers and manufacturers to rapidly and inexpensively test the parts of the design that are most likely to have problems, solve those problems, and then build the full design.

This counter-intuitive idea —that the quickest way to build something is, first to build something else— is shared by scaffolding and the telescope rule.

Modern trends

With the recent advances in computer modeling it is becoming practical to eliminate the creation of a physical prototype (except possibly at greatly reduced scales for promotional purposes), instead modeling all aspects of the final product as a computer model. An example of such a development can be seen in the Boeing 787 Dreamliner, in which the first full sized physical realization is made on the series production line. Computer modeling is now being extensively used in automotive design, both for form (in the styling and aerodynamics of the vehicle) and in function — especially for improving vehicle crashworthiness and in weight reduction to improve mileage.

Mechanical and electrical engineering

The most common use of the word prototype is a functional, although experimental, version of a non-military machine (e.g., automobiles, domestic appliances, consumer electronics) whose designers would like to have built by mass production means, as opposed to a mockup, which is an inert representation of a machine's appearance, often made of some non-durable substance.

An electronics designer often builds the first prototype from breadboard or stripboard or perfboard, typically using "DIP" packages.

However, more and more often the first functional prototype is built on a "prototype PCB" almost identical to the production PCB, as PCB manufacturing prices fall and as many components are not available in DIP packages, but only available in SMT packages optimized for placing on a PCB.

A prototype of the Polish economy hatchback car Beskid 106 designed in the 1980s.

Builders of military machines and aviation prefer the terms "experimental" and "service test".

Electronics prototyping

In electronics, prototyping means building an actual circuit to a theoretical design to verify that it works, and to provide a physical platform for debugging it if it does not. The prototype is often constructed using techniques such as wire wrap or using veroboard or breadboard, that create an electrically correct circuit, but one that is not physically identical to the final product.

Open-source tools exist to document electronic prototypes (especially the breadboard-based ones) and move forward toward production such as Fritzing and Arduino.

A technician can build a prototype (and make additions and modifications) much more quickly with these techniques —however, it is much faster and usually cheaper to mass produce custom printed circuit boards than these other kinds of prototype boards. This is for the same reasons that writing a poem is fastest by hand for one or two, but faster by printing press if you need several thousand copies.

The proliferation of quick-turn pcb fab companies and quick-turn pcb assembly houses has enabled the concepts of rapid prototyping to be applied to electronic circuit design. It is now possible, even with the smallest passive components and largest fine-pitch packages, to have boards fabbed and parts assembled in a matter of days.

Computer programming/computer science

In many programming languages, a *function prototype* is the declaration of a subroutine or function. (This term is rather C/C++-specific; other terms for this notion are *signature*, *type* and *interface*.) In prototype-based programming (a form of object-oriented programming), new objects are produced by cloning existing objects, which are called prototypes.[3]

The term may also refer to the Prototype Javascript Framework.

Additionally, the term may refer to the prototype design pattern.

Prototype software is often referred to as alpha grade, meaning it is the first version to run. Often only a few functions are implemented, the primary focus of the alpha is to have a functional base code on to which features may be added. Once alpha grade software has most of the required features integrated into it, it becomes beta software for testing of the entire software and to adjust the program to respond correctly during situations unforeseen during development.[4]

Often the end users may not be able to provide a complete set of application objectives, detailed input, processing, or output requirements in the initial stage. After the user evaluation, another prototype will be built based on feedback from users, and again the cycle returns to customer evaluation. The cycle starts by listening to the user, followed by building or revising a mock-up, and letting the user test the mock-up, then back. There is now a new generation of tools called Application Simulation Software which help quickly simulate application before their development.

Extreme programming uses iterative design to gradually add one feature at a time to the initial prototype.

Continuous learning approaches within organizations or businesses may also use the concept of business or process prototypes through software models.

Data prototyping

A *data prototype* is a form of *functional* or *working* prototype. The justification for its creation is usually a data migration, data integration or application implementation project and the raw materials used as input are an instance of all the relevant data which exists at the start of the project.

The objectives of *data prototyping* are to produce:

- A set of data cleansing and transformation rules which have been *seen* to produce data which is all fit for purpose.
- A dataset which is the result of those rules being applied to an instance of the relevant raw (source) data.

To achieve this, a data architect uses a graphical interface to interactively develop and execute transformation and cleansing rules using raw data. The resultant data is then evaluated and the rules refined. Beyond the obvious visual checking of the data *on-screen* by the data architect, the usual evaluation and validation approaches are to use Data profiling software and then to insert the resultant data into a test version of the target application and trial its use.

Scale modeling

In the field of scale modeling (which includes model railroading, vehicle modeling, airplane modeling, military modeling, etc.), a prototype is the real-world basis or source for a scale model—such as the real EMD GP38-2 locomotive—which is the prototype of Athearn's (among other manufacturers) locomotive model. Technically, any non-living object can serve as a prototype for a model, including structures, equipment, and appliances, and so on, but generally prototypes have come to mean full-size real-world vehicles including automobiles (the prototype 1957 Chevy has spawned many models), military equipment (such as M4 Shermans, a favorite among US Military modelers), railroad equipment, motor trucks, motorcycles, and space-ships (real-world such as Apollo/Saturn Vs, or the ISS).

There is debate whether 'fictional' or imaginary items can be considered prototypes (such as Star Wars or Star Trek starships, since the feature ships themselves *are* models or CGI-artifacts); however, humans and other living items are never called prototypes, even when they are the basis for models and dolls (especially - action figures).

As of 2005, conventional rapid prototype machines cost around £25,000.[5]

Metrology

In the science and practice of metrology, a **prototype** is a human-made object that is used as *the* standard of measurement of some physical quantity to base all measurement of that physical quantity against. Sometimes this standard object is called an **artifact**. In the International System of Units (**SI**), the only prototype remaining in current use is the International Prototype Kilogram, a solid platinum-iridium cylinder kept at the Bureau International des Poids et Mesures (International Bureau of Weights and Measures) in Sèvres France (a suburb of Paris) that by definition is the mass of exactly one kilogram. Copies of this prototype are fashioned and issued to many nations to represent the national standard of the kilogram and are periodically compared to the Paris prototype.

Until 1960, the meter was defined by a platinum-iridium prototype bar with two scratch marks on it (that were, by definition, spaced apart by one meter), the International Prototype Metre, and in 1983 the meter was redefined to be the distance in free space covered by light in 1/299,792,458 of a second (thus *defining* the speed of light to be 299,792,458 meters per second).

It is widely believed that the kilogram prototype standard will be replaced by a definition of the kilogram that will define another physical constant (likely either Planck's constant or the elementary charge) to a defined numerical value, thus obviating the need for the prototype and removing the possibility of the prototype (and thus the standard and definition of the kilogram) changing very slightly over the years because of loss or gain of atoms.

Sciences

In many sciences, from pathology to taxonomy, prototype refers to a disease, species, etc. which sets a good example for the whole category. For example, the vaccina virus and Senegal bichir are regarded as the prototypes of their respective species.

Advantages and disadvantages

Advantages of prototyping

- May provide the proof of concept necessary to attract funding
- Early visibility of the prototype gives users an idea of what the final system looks like
- Encourages active participation among users and producer
- Enables a higher output for user
- Cost effective (Development costs reduced).
- Increases system development speed
- Assists to identify any problems with the efficacy of earlier design, requirements analysis and coding activities
- Helps to refine the potential risks associated with the delivery of the system being developed
- Various aspects can be tested and quicker feedback can be got from the user
- Helps to deliver the product in quality easily
- User interaction available during development cycle of prototype

Disadvantages of prototyping

- Producer might produce a system inadequate for overall organization needs
- User can get too involved whereas the program can not be to a high standard
- Structure of system can be damaged since many changes could be made
- Producer might get too attached to it (might cause legal involvement)
- Not suitable for large applications
- Over long periods, can cause loss in consumer interest and subsequent cancellation due to a lack of a market (for commercial products)
- May slow the development process, if there are large number of end users to satisfy.

References

[1] "Prototyping Definition" (http://www.pcmag.com/encyclopedia_term/0,1233,t=prototyping&i=49886,00.asp). *PC Magazine*. . Retrieved 2012-05-03.

[2] Online Etymology Dictionary (http://www.etymonline.com/index.php?search=prototype&searchmode=none)

[3] "5.5 Function Prototypes" (http://h30097.www3.hp.com/docs/base_doc/DOCUMENTATION/V40F_HTML/AQTLTBTE/ DOCU_055.HTM). *HP*. . Retrieved 2012-05-03.

[4] "Alpha Version Definition" (http://www.pcmag.com/encyclopedia_term/0,1233,t=alpha+version&i=37675,00.asp). *PC Magazine*. . Retrieved 2012-05-03.

[5] Bath.ac.uk (http://www.bath.ac.uk/pr/releases/replicating-machines.htm)

Military_transport_aircraft

Military transport aircraft are typically fixed and rotary wing cargo aircraft which are used to deliver troops, weapons and other military equipment by a variety of methods to any area of military operations around the surface of the planet, usually outside of the commercial flight routes in uncontrolled airspace. Originally derived from bombers, military transport aircraft were used for delivering airborne forces during the Second World War and towing military gliders. Some military transport aircraft are tasked to performs multi-role duties such as aerial refueling and, tactical, operational and strategic airlifts onto unprepared runways, or those constructed by engineers.

C-17 Globemaster III transport aircraft

Fixed wing transport, tanker, cargo airlift, and utility aircraft

Fixed wing transport aircraft are largely defined in terms of their range capability as strategic airlift, airlift and tactical airlift to reflect the needs of the land forces which they most often support. These roughly correspond to the commercial flight length distinctions:

Short-haul flight:	<3 hours
Medium-haul flight:	3 to 6 hours
Long-haul flight:	>6 hours

A more specialised role of a cargo aircraft is that of transporting fuel in support of other aircraft with more limited flight endurance such as fighters or helicopters. Smaller cargo aircraft, known as "utility", are often used to transport military communications equipment as temporary or permanent platforms, and in the command role by providing airborne command post or as an air ambulance.

Active military airlift transports and cargo aircraft

The C-5 Galaxy

The Airbus A400M

Kawasaki XC-2

The Il-76 'Candid'

The Transall C160NG, Escadron Anjou, French
Air Force

Manufacturer	Model	first flight	max Payload (t)	Cruise (km/h)	max range (km)	MTOW
Airbus	A330 MRTT	2007	45	860	14,800	223
Airbus	A400M	2009	37	780	9,300	141
Alenia	C-27J Spartan	2008	11.5	583	5,926	31.8
Antonov	An-12	1957	20	777	5,700	61
Antonov	An-22 Antei	1965	80	740	5,000	250
Antonov	An-26	1969	5.5	440	2,550	24
Antonov	An-32	1976	6.7	480	2,500	26.9
Antonov	An-70	1994	47	729	6,600	145
Antonov	An-72	1977	7.5	600	4,800	33
Antonov	An-124 Ruslan	1982	150	800-850	5,410	405
Antonov	An-225 Mriya	1988	250	800	15,400	600
AVIC	Y-9	2008	25	650	7,800	77
Bell/Boeing	V-22 Osprey	1989	6.8	396	1,627	27.4
Boeing	C-17 Globemaster III	1991	77.5	830	4,482	265
CASA	C-212 Aviocar	1971	2.8	315	1,433	8
CASA/Indonesian Aerospace	CN-235	1983	5	509	5,003	15.1
CASA	C-295	1998	9.3	481	5,630	23.2
de Havilland Canada	C-7 Caribou	1958	3.6	348	2,103	14.2
Douglas	C-47	1943	3	360	2,600	10.5
Grumman	C-1 Trader	1952	1.6	462	2,092	13.2

Grumman	C-2 Greyhound	1964	4.5	465	2,400	24.7
Embraer	C-390	2014	19	900	6,200	72
Fairchild	C-123 Provider	1949	11	367	1,666	27
Ilyushin	Il-76	1971	47	900	4,400	210
Ilyushin	Il-112	2011	5.9	550	5,000	20
Kawasaki	C-1	1970	11.9	657	1,300	45
Kawasaki	XC-2	2010	37.6	890	6,500	120
Lockheed	C-5 Galaxy	1968	122	907	4,445	381
Lockheed	C-130 Hercules	1954	20	540	3,800	70.3
Lockheed	C-141 Starlifter	1963	45	912	9,880	147
Short Brothers	C-23 Sherpa	1982	3.2	296	1,239	3.2
Transport Allianz	Transall C-160	1963	16	513	1,850	49.2
UAC and HAL	UAC/HAL Multirole Transport Aircraft	2015	22	830	2,500	68

Commercial and tanker aircraft

Commercial aircraft used in military role   **Military aerial refuelling tankers**

- C-9 Skytrain II
- C-12 Huron
- C-20 Gulfstream III
- C-21 Learjet
- C-32
- C-40 Clipper
- VC-25A *Air Force One*

RAF TriStar refuelling US Navy F/A-18s

- Airbus A400M
- Airbus A310 MRTT
- Airbus A330 MRTT
- C-135 Stratolifter
- Lockheed TriStar
- KC-10 Extender
- KC-135 Stratotanker
- Vickers VC10
- Ilyushin Il-78

Transport helicopters

Military transport helicopters are used in places where the use of conventional aircraft is impossible. For example the military transport helicopter is the primary transport asset of US Marines deploying from LHDs and LHA. The landing possibilities of helicopter are almost unlimited, and where landing is impossible, for example densely packed jungle, the ability of the helicopter to hover allows troops to deploy by abseiling and roping.

Kazak Mi-17 helicopter, some members of the Mil Mi-8 family can carry both weapons and troops

Transport helicopters are operated in assault, medium and heavy classes. Air assault helicopters are usually the smallest of the transport types, and designed to move an infantry section and their equipment. Helicopters in the assault role are generally armed for self protection both in transit and for suppression of the landing zone. This armament may be in the form of door gunners, or the modification of the helicopter with stub wings and pylons for the carriage of missiles and rocket pods. For example the Sikorsky S-70 fitted with the ESSM (External Stores Support System) and the *Hip E* variant of the Mil Mi-8 can carry as much disposable armament as some dedicated attack helicopters. The assault helicopter can be thought of as the modern successor to the military glider.

Not all militaries are able to operate a full range of transport helicopters so the medium transport type as the most useful compromise is probably the most common specialist transport type. Medium transport helicopters are generally capable of moving up to a platoon of infantry and are capable of being able to transport towed artillery or light vehicles either internally or as under-slung roles. Unlike the assault helicopter they are usually not expected to land directly in a contested landing zone, but are used to reinforce and resupply landing zones taken by the initial assault wave. Examples include the unarmed versions of the Mil Mi-8, Super Puma, and CH-46 Sea Knight.

CH-53 with possible internal load

Heavy lift helicopters are the largest and most capable of the transport types, currently limited in service to the CH-53 Sea Stallion and related CH-53E Super Stallion, CH-47 Chinook, Mil Mi-26, and Aérospatiale Super Frelon. Capable of lifting up to 80 troops and moving small

CH-54B carrying an M551 Sheridan tank

AFVs (usually as slung loads but also internally), these helicopters operate in the tactical transport role in much the same way as small fixed wing turboprop air-lifters. The lower speed, range and increased fuel consumption of helicopters are more than compensated by their ability to operate virtually anywhere.

See also

- Cargo aircraft
- Loadmaster
- Airlift

Monoplane

For Félix du Temple's invention, see Monoplane (1874)

Monoplane

The low-wing of a Curtiss P-40

A **monoplane** is a fixed-wing aircraft with one main set of wing surfaces, in contrast to a biplane or triplane. Since the late 1930s it has been the most common form for a fixed wing aircraft.

Types of monoplane

The main distinction between types of monoplane is where the wings attach to the fuselage:

- **low-wing**, the wing lower surface is level with (or below) the bottom of the fuselage
- **mid-wing**, the wing is mounted mid-way up the fuselage
- **shoulder wing**, the wing is mounted above the fuselage middle
- **high-wing**, the wing upper surface is level with or above the top of the fuselage
- **parasol-wing**, the wing is located above the fuselage and is not directly connected to it, structural support being typically provided by a system of struts, and, especially in the case of older aircraft, wire bracing.

The mid-wing of a de Havilland Vampire T11.

The shoulder-wing of an ARV Super2.

History

Most of the first attempts of heavier-than-air flying machines were monoplanes. Notably, the *Monoplane* built in 1874 by Felix du Temple de la Croix, a large aircraft made of aluminium with a wingspan of 13 m (**unknown operator: u'strong'** ft) and a weight of only 80 kg (**unknown operator: u'strong'** lb) (without the pilot). Several trials were made with the plane in Brest, France, and it is generally recognized that it achieved lift off under its own power after a ski-jump run, glided for a short time and returned safely to the ground, possibly making it the first successful powered flight in history, depending on the definition — since the flight was only a few feet high, and was not truly under control.

Other early attempts of flight by a monoplane were carried out in 1884 by Alexander Mozhaysky.[1]

The first successful aircraft were biplanes, but many pioneering aircraft were monoplanes, for instance Blériot XI that flew across the English Channel in 1909. Throughout 1909-1910 Hubert Latham set multiple altitude records in his Antoinette IV monoplane, initially achieving 155 m (**unknown operator: u'strong'** ft) then raising it to 1384 m (**unknown operator: u'strong'** ft).[2] The Fokker Eindecker of 1915 was a successful fighter aircraft. The Junkers J 1 was an early German "technology demonstrator" monoplane, and the world's very first practical all-metal aircraft of any type to fly, with the J 1's first flight occurring in December 1915.

According to the Encyclopædia Britannica, the first successful monoplane flight — a duration of 12 m (40 feet) at Montesson, France — was on March 18, 1906 in a craft fashioned by Traian Vuia, a Romanian inventor.[3] [4]

Nonetheless, relatively few monoplane types were built between 1914, and the late 1920s, compared with the number of biplanes. The reasons for this were primarily structural. In the days when wings (whether biplane or monoplane) were thin, lightly built structures, braced by struts, steel wire or cables - the biplane wing formed a strong and fairly rigid lattice truss structure, in which the two wing surfaces were braced against each other. Early monoplane wings, on the other hand, tended to be liable to twist under aerodynamic loads, rendering proper lateral control very difficult. They were also much more liable to breakage in flight.

Once all-metal construction and the cantilever wing, both having been pioneered by Hugo Junkers in 1915, became common after World War I's end, however, the day of the biplane very quickly passed, and the monoplane became the usual configuration for a fixed-wing aircraft.

Most military aircraft of WW2 were monoplanes, as have been virtually all aircraft since. Thus the superiority of the design over the biplane was established, and biplanes have been relegated to specialized applications ever since.[3]

The high-wing of a de Havilland Canada Dash 8.

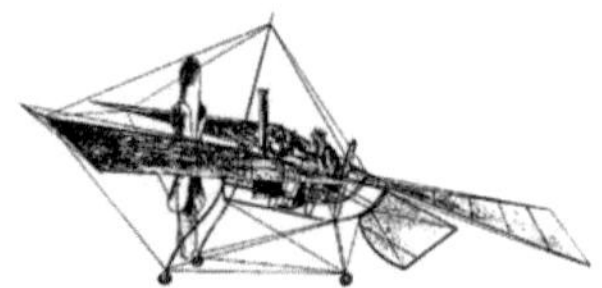

Félix du Temple's 1874 *Monoplane*.

A parasol wing Pietenpol Air Camper amateur-built aircraft.

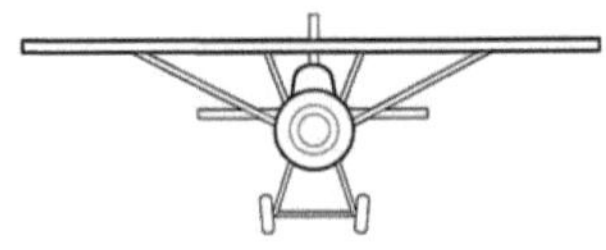

Schematic head-on illustration of a parasol wing

See also

- Biplane
- Triplane

References

[1] Gray, Carroll (undated). "Aleksandr Fyodorovich Mozhaiski" (http://www.flyingmachines.org/moz.html). . Retrieved 8 November 2011.
[2] King, *Windkiller*, p. 227.
[3] "Monoplane" (http://www.britannica.com/EBchecked/topic/390020/monoplane). *Encyclopædia Britannica, Encyclopædia Britannica Online*. Encyclopædia Britannica Inc. . Retrieved March 18, 2012.
[4] Compare, Trajan Vuia, (1872-1950) (http://www.ctie.monash.edu.au/hargrave/vuia.html)]

Undercarriage

Undercarriage

Undercarriage of a Boeing 777-300

The **undercarriage** or **landing gear** in aviation, is the structure that supports an aircraft on the ground and allows it to taxi, takeoff and land. Typically wheels are used, but skids, skis, floats or a combination of these and other elements can be deployed, depending on the surface.

Undercarriage of an Airbus A380

Overview

Landing gear usually includes wheels equipped with shock absorbers for solid ground, but some aircraft are equipped with skis for snow or floats for water, and/or skids or pontoons (helicopters).

The undercarriage is a relatively heavy part of the vehicle, it can be as much as 7% of the takeoff weight, but more typically is 4-5%.[1]

Gear arrangements

A SAN Jodel D.140 Mousquetaire with conventional "taildragger" undercarriage.

Wheeled undercarriages normally come in two types: *conventional or "taildragger" undercarriage*, where there are two main wheels towards the front of the aircraft and a single, much smaller, wheel or skid at the rear; or *tricycle undercarriage* where there are two main wheels (or wheel assemblies) under the wings and a third smaller wheel in the nose. The taildragger arrangement was common during the early propeller era, as it allows more room for propeller clearance. Most modern aircraft have tricycle undercarriages. Taildraggers are considered harder to land and take off (because the arrangement is *unstable*, that is, a small deviation from straight-line travel is naturally amplified by the greater drag of the mainwheel which has moved farther away from the plane's centre of gravity due to the deviation), and usually require special pilot training. Sometimes a small tail wheel or skid is added to aircraft with tricycle undercarriage, in case of tail strikes during take-off. The Concorde, for instance, had a retractable tail "bumper" wheel, as delta winged aircraft need a high angle when taking off. The Boeing 727 also had a retractable tail bumper. Some aircraft with retractable conventional landing gear have a fixed tailwheel, which generates minimal drag (since most of the airflow past the tailwheel has been blanketed by the fuselage) and even improves yaw stability in some cases.

A Mooney M20J with tricycle undercarriage.

Retractable gear

To decrease drag in flight some undercarriages retract into the wings and/or fuselage with wheels flush against the surface or concealed behind doors; this is called *retractable gear*.

If the wheels rest protruding and partially exposed to the air stream after being retracted, the system is called semi-retractable.

Main and nosewheel undercarriage of an Airbus A330

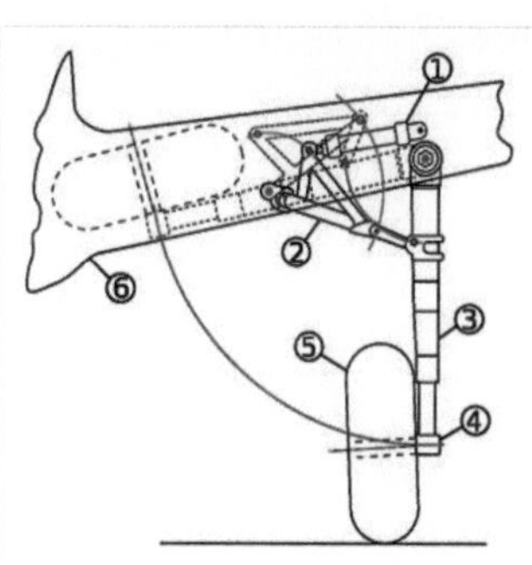
Schematic showing hydraulically operated landing gear, with the wheel stowed into the wing root of the aircraft

Most retraction systems are hydraulically operated, though some are electrically operated or even manually operated. This adds weight and complexity to the design. In retractable gear systems, the compartment where the wheels are stowed are called wheel wells, which may also diminish valuable cargo or fuel space.

A design for retractable landing gear was first seen in 1876 in plans for an amphibious monoplane designed by Frenchmen Alphonse Pénaud and Paul Gauchot. Aircraft with at least partially retractable landing gear did not appear until 1917, and it was not until the late 1920s and early 1930s that such aircraft became common, with Grover Loening's military aircraft designs being among the first routinely using them for the main undercarriage members, in a system later licensed and used by his friend Leroy Grumman's aviation firm. By then, aircraft performance was improved to the point where the aerodynamic advantage of a retractable undercarriage justified the added complexity, weight and interior space penalties. An alternate method of reducing the aerodynamic penalty imposed by fixed undercarriage is to attach aerodynamic fairings (often called "spats" or "pants") on the undercarriage, with only the bottoms of the wheels exposed.

A Boeing 737-700 with main undercarriage retracted in the wheel wells without landing gear doors

Pilots confirming that their landing gear is down and locked refer to "three green" or "three in the green.", a reference to electrical indicator lights from the nosewheel and the two main gears. Amber lights indicate the gears are in the up-locked position; red lights indicates that the landing gear is in transit (neither down and locked nor fully retracted).[2]

Multiple redundancies are usually provided to prevent a single failure from failing the entire landing gear extension process. Whether electrically or hydraulically operated, the landing gear can usually be powered from multiple sources. In case the power system fails, an emergency extension system is always available. This may take the form of a manually operated crank or pump, or a mechanical free-fall mechanism which disengages the uplocks and allows the landing gear to fall due to gravity. Some high-performance aircraft may even feature a pressurized-nitrogen back-up system.

Large aircraft

As aircraft grow larger, they employ more wheels to cope with the increasing weights. The earliest "giant" aircraft ever placed in quantity production, the Zeppelin-Staaken R.VI German World War I long-range bomber of 1916, used a total of eighteen wheels for its undercarriage, split between two wheels on its nose gear struts, and a total of sixteen wheels on its main gear units under each tandem engine nacelle, to support its loaded weight of almost 12 metric tons. Later, during World War II, the experimental German Arado Ar 232 cargo aircraft used a centerline row of ten "twinned" fixed wheel sets directly under the fuselage centreline to handle heavier loads while on the ground, as the earliest known example of multiple "tandem wheels" on an aircraft, like many of today's large cargo aircraft use for their retractable main gear setups (usually mounted on the lower corners of the central fuselage structure). The Airbus A340-500/-600 has an additional four-wheel undercarriage bogie on the fuselage centreline, much like the twin-wheel unit in the same general location, used on later DC-10 and MD-11 airliners. The Boeing 747 has five sets of wheels: a nose-wheel assembly and four sets of four-wheel bogies. A

Tire Arrangements of large aircraft

Main landing gear on an Antonov An-225

set is located under each wing, and two inner sets are located in the fuselage, a little rearward of the outer bogies, adding up to a total of eighteen wheels and tires. The Airbus A380 also has a four-wheel bogie under each wing with two sets of six-wheel bogies under the fuselage. The enormous Ukrainian Antonov An-225 jet cargo aircraft has one of the largest, if not the largest, number of individual wheel/tire assemblies in its landing gear design - with a total of four wheels on the twin-strut nose gear units, and a total of 28 main gear wheel/tire units, adding up to a total of 32 wheels and tires.

An Airbus A340-600, which has an undercarriage on the fuselage belly in addition to the wings

Unusual types of gear

Rarely, planes use wheels only for takeoff and drop them afterwards, to gain the improved streamlining without the complexity, weight and space requirements of a retraction mechanism, with such jettisonable wheels sometimes mounted onto axles that were part of a separate "dolly" (for main wheels only) or "trolley" (for a three wheel set with a nosewheel) chassis. In this case, landing is achieved on skids or similar simple devices. Historical examples include the "dolly"-using Messerschmitt Me 163 rocket fighter, the Messerschmitt Me 321 *Gigant* troop glider, and the first eight "trolley"-using prototypes of the Arado Ar 234 jet reconnaissance bomber. The main disadvantage to using the takeoff dolly/trolley and landing skid(s) system on German World War II aircraft, was that aircraft would likely be scattered all

An Me 163B *Komet* with its two-wheel take-off "dolly" in place

over a military airfield after they had landed from a mission, and would be unable to taxi on their own to an appropriately hidden "dispersal" location, which could easily leave them vulnerable to being shot up by attacking Allied fighters. A related contemporary example are the wingtip support wheels ("Pogos") on the Lockheed U-2 reconnaissance aircraft, which fall away after take-off and drop to earth; the aircraft then relies on titanium skids on the wingtips for landing.

Some main gear struts on World War II aircraft, in order to allow a single-leg main gear to more efficiently store the wheel within either the wing or an engine nacelle, rotated the single gear strut through a 90° angle during the rearwards-retraction sequence to allow the main wheel to rest "flat" above the lower end of the main gear strut, or flush within the wing, when fully retracted. Examples are the Curtiss P-40, Vought F4U Corsair, Grumman F6F Hellcat, Messerschmitt Me 210 and Junkers Ju 88. The Aero Commander family of twin-engined

Hawker Siddeley Harrier GR7 (ZG472). The two mainwheels are in line astern under the fuselage, with a smaller wheel on each wing

business aircraft also shares this feature on the main gears, which retract aft into the ends of the engine nacelles. The rearward-retracting nosewheel strut on the Heinkel He 219 and the forward-retracting nose gear strut on the later Cessna Skymaster similarly rotated 90 degrees as they retracted.

A Royal Air Force P-47 with its raked-forward main gear, and rearward-angled main wheel position (when retracted) indicated by the just-visible open wheel door.

On most World War II single-engined fighter aircraft (and even one German heavy bomber design) with sideways retracting main gear, the main gear that retracted into the wings was meant to be raked forward, towards the aircraft's nose in the "down" position for better ground handling, with a retracted position that placed the main wheels at some angle "behind" the main gear's attachment point to the airframe - this led to a complex geometry for setting up the angles for the retraction mechanism's axis of rotation, with some aircraft, like the P-47 Thunderbolt, even mandating that the main gear struts lengthen as they were extended down from the wings to assure proper ground clearance for its large four-bladed propeller. One exception to the need for this complexity in many WW II fighter aircraft was Japan's famous Zero fighter, whose main gear stayed at a perpendicular angle to the centreline of the aircraft when extended, as seen from the side.

An unusual undercarriage configuration is found on the Hawker Siddeley Harrier, which has two mainwheels in line astern under the fuselage (called a bicycle or *tandem* layout) and a smaller wheel near the tip of each wing. On second generation Harriers, the wing is extended past the outrigger wheels to allow greater wing-mounted munition loads to be carried.

Experimental tracked gear on a B-36 Peacemaker

A multiple tandem layout was used on some military jet aircraft during the 1950s, pioneered by the Martin XB-51, and later used on such aircraft as the U-2, Myasishchev M-4, Yakovlev Yak-25, Yak-28 and the B-47 Stratojet because it allows room for a large internal bay between the main wheels. A variation of the multi tandem layout is also used on the B-52 Stratofortress which has four main wheel bogies (two forward and two aft) underneath the fuselage and a small outrigger wheel supporting each wing-tip. The B-52's landing gear is also unique in that all four pairs of main wheels can be steered. This allows the landing gear to line up with the runway and thus makes crosswind landings easier (using a technique called *crab landing*). The challenge of designing a tandem-gear layout is that the aircraft has to sit (on the ground) at the optimum flight angle for landing - when the plane is nearly in a stalled attitude just before touchdown, both fore and aft wheels must be ready to contact the runway. Otherwise there will be a vicious jolt as the higher wheel falls to the runway at the stall.

One very early undercarriage arrangement that passively allowed for castoring during crosswind landings, unlike the "active" arrangement on the B-52, was pioneered on the Bleriot VIII design of 1908. It was later used in the much more famous Blériot XI Channel-crossing aircraft of 1909 and also copied in the earliest examples of the Etrich Taube. In this arrangement the main landing gear's shock absorption was taken up by a vertically sliding bungee cord-sprung upper member. The vertical post along which the upper member slid to take landing shocks also had its lower end as the rotation point for the forward end of the main wheel's suspension fork, allowing the main gear to pivot on moderate crosswind landings.

The "castoring" main gear arrangement on a Blériot XI

Light aircraft

For light aircraft a type of landing gear which is economical to produce is a simple wooden arch laminated from ash, as used on some homebuilt aircraft. A similar arched gear is often formed from spring steel. The Cessna Airmaster was among the first aircraft to use spring steel landing gear. The main advantage of such gear is that no other shock-absorbing device is needed; the deflecting leaf provides the shock absorption.

Gliders

To minimize drag, modern gliders most usually have a single wheel, retractable or fixed, centered under the fuselage, which is referred to as *monowheel gear* or *monowheel landing gear*. Monowheel gear is also used on some powered aircraft, where drag reduction is a priority, such as the Europa XS. Some gliders from prior to the Second World War used a take-off dolly that was jettisoned on take-off and then landed on a fixed skid.[3]

Monowheel landing gear on a Schleicher ASK 21 glider

Steering

There are several types of steering. Taildragger aircraft may be steered by rudder alone (depending upon the prop wash produced by the aircraft to turn it) with a freely pivoting tail wheel, or by a steering linkage with the tail wheel, or by *differential braking* (the use of independent brakes on opposite sides of the aircraft to turn the aircraft by slowing one side more sharply than the other). Aircraft with tricycle landing gear usually have a steering linkage with the nose wheel (especially in large aircraft), but some allow the nose wheel to pivot freely and use differential braking and/or the rudder to steer the aircraft.

Two mechanics replacing a tire on a P-3C Orion

Some aircraft require that the pilot steer by using rudder pedals; others allow steering with the yoke or control stick. Some allow both. Still others have a separate control, called a *tiller*, used for steering on the ground exclusively.

Rudder steering

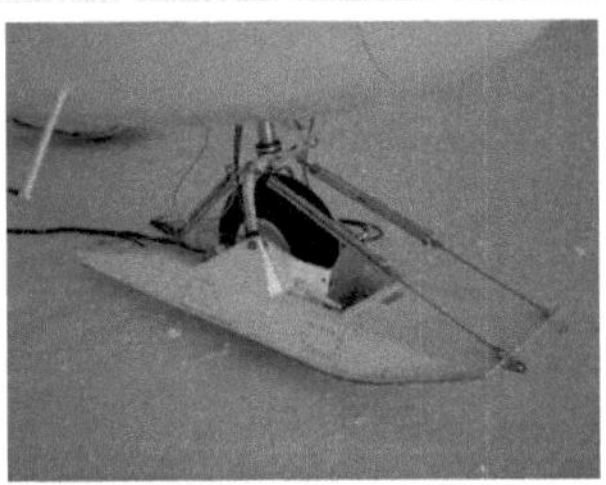
Wheel-skis

When an aircraft is steered on the ground exclusively using the rudder, turning the plane requires that a substantial airflow be moving past the rudder, which can be generated either by the forward motion of the aircraft or by thrust provided by the engines. Rudder steering requires considerable practice to use effectively. Although it requires air movement, it has the advantage of being independent of the landing gear, which makes it useful for aircraft equipped with fixed floats or skis.

Direct steering

Some aircraft link the yoke, control stick, or rudder directly to the wheel used for steering. Manipulating these controls turns the steering wheel (the nose wheel for tricycle landing gear, and the tail wheel for taildraggers). The connection may be a firm one in which any movement of the controls turns the steering wheel (and vice versa), or it may be a soft one in which a spring-like mechanism twists the steering wheel but does not force it to turn. The former provides positive steering but makes it easier to skid the steering wheel; the latter provides softer steering (making it easy to overcontrol) but reduces the probability of skidding. Aircraft with retractable gear may disable the steering mechanism wholly or partially when the gear is retracted.

Differential braking

Differential braking depends on asymmetric application of the brakes on the main gear wheels to turn the aircraft. For this, the aircraft must be equipped with separate controls for the right and left brakes (usually on the rudder pedals). The nose or tail wheel usually is not equipped with brakes. Differential braking requires considerable skill. In aircraft with several methods of steering that include differential braking, differential braking may be avoided because of the wear it puts on the braking mechanisms. Differential braking has the advantage of being largely independent of any movement or skidding of the nose or tail wheel.

Wing and fuselage undercarriages on a Boeing 747-400, shortly before landing

Tiller steering

A tiller in an aircraft is a small wheel or lever, sometimes accessible to one pilot and sometimes duplicated for both pilots, that controls the steering of the aircraft while it is on the ground. The tiller may be designed to work in combination with other controls such as the rudder or yoke. In large airliners, for example, the tiller is often used as the sole means of steering during taxi, and then the rudder is used to steer during take-off and landing, so that both aerodynamic control surfaces and the landing gear can be controlled simultaneously when the aircraft is moving at aerodynamic rates of speed.

Landing gear and accidents

Malfunctions or human errors (or a combination of these) related to retractable landing gear have been the cause of numerous accidents and incidents throughout aviation history. Distraction and preoccupation during the landing sequence played a prominent role in the approximately 100 gear-up landing incidents that occurred each year in the United States between 1998 and 2003.[4] A gear-up landing incident, also known as a belly landing, is an accident that may result from the pilot simply forgetting, or failing, to lower the landing gear before landing or a mechanical malfunction that does not allow the landing gear to be lowered. Although rarely fatal, a gear-up landing is very expensive, as it causes massive airframe damage. For propeller driven aircraft it almost always requires a complete rebuild of engines

JetBlue Airways Flight 292, an Airbus A320, making an emergency landing on runway 25L at LAX in 2005 after the front landing gear malfunctioned

because the propellers strike the ground and suffer a sudden stoppage if they are running during the impact. Many aircraft between the wars - at the time when retractable gear was becoming commonplace - were deliberately designed to allow the bottom of the wheels to protrude below the fuselage even when retracted to reduce the damage caused if the pilot forgot to extend the landing gear or in case the plane was shot down and forced to crash-land. Examples include the Avro Anson, Boeing B-17 Flying Fortress and the Douglas DC-3. The modern-day Fairchild-Republic A-10 Thunderbolt II carries on this legacy: it is similarly designed in an effort to avoid (further) damage during a gear-up landing, a possible consequence of battle damage.

Some aircraft have a stiffened fuselage bottom or added firm structures, designed to minimise structural damage in a wheels-up landing. When the Cessna Skymaster was converted for a military spotting role (the O-2 Skymaster), fiberglass railings were added to the length of the fuselage; they were adequate to support the aircraft without damage if it was landed on a grassy surface.

The Bombardier Dash 8 is notorious for its landing gear problems. There were three incidents involved, all of them involving Scandinavian Airlines, flights SK1209, SK2478, and SK2867. This led to Scandinavian retiring all of its Dash 8s. The cause of these incidents was a locking mechanism that failed to work properly. This also caused concern for the aircraft for many other airlines that found similar problems, Bombardier Aerospace ordered all Dash 8s with 10,000 or more hours to be grounded, it was soon found that 19 Horizon Airlines Dash 8s had locking mechanism problems, so did 8 Austrian Airlines planes, this did cause several hundred flights to be canceled.

On September 21, 2005, JetBlue Airways Flight 292 successfully landed with its nose gear turned 90 degrees sideways, resulting in a shower of sparks and flame after touchdown. This type of incident is very uncommon as the nose oleo struts are designed with centering cams to hold the nosewheels straight until they are compressed by the weight of the aircraft.

On November 1, 2011, LOT Polish Airlines Flight LO16 successfully belly landed at Warsaw Chopin Airport due to technical failures, all 231 people on board escaped with no injury.[5]

Automatic extension systems

The Piper Arrow was originally fitted with a system that automatically extended the landing gear when certain power and flap settings were selected. The manufacturer issued an Airworthiness Directive for owners to disable this system. Pilots were found to be relying on this system to extend the gear in routine flight operations, rather than just as an emergency backup. If the gear failed to extend then the manufacturer was exposed to liability for the resulting gear-up landing. There were also concerns over unintentional gear extension incidents where pilots placed the aircraft in "bad-weather" (low-power setting, flaps down) configuration and inadvertently activated the gear extension system.

Emergency extension systems

In the event of a failure of the aircraft's landing gear extension mechanism a back-up is provided. This may be an alternate hydraulic system, a hand-crank, compressed air (nitrogen), pyrotechnic or free-fall system.[6]

A free-fall or gravity drop system uses gravity to deploy the landing gear into the down and locked position. To accomplish this the pilot activates a switch or mechanical handle in the cockpit, which releases the up-lock. Gravity then pulls the landing gear down and deploys it. Once in position the landing gear is mechanically locked and safe to use and land on.[7] [8]

Stowaways

For main article and stowaway accidents, see Stowaway.

Unauthorized passengers have been known to stowaway on larger aircraft by climbing a landing gear strut and riding within the compartment. There are extreme dangers to this practice.

See also

- Landing gear extender
- Undercarriage arrangements of jetliners and other aircraft.
- Dayton-Wright Racer, an early example of an airplane with retractable landing gear.
- Verville Racer Aircraft, an early example of an airplane with retractable landing gear.

References

[1] Aircraft Design By Ajoy Kumar Kundu (http://books.google.co.uk/books?id=NeHoahlhCGMC&pg=PA194&lpg=PA194& dq=undercarriage+mass&source=bl&ots=DQbaVHo4y5&sig=vRHe2ek6Pft8m8om9Q7IKnB-zBI&hl=en& ei=u6N_TvaYL8Op8QPFv-CpAQ&sa=X&oi=book_result&ct=result&resnum=5&ved=0CF0Q6AEwBA#v=onepage&q=undercarriage mass&f=false)

[2] Retractable Landing Gear (http://www.flightsimbooks.com/flightsimhandbook/CHAPTER_02_10_Retractable_Landing_Gear.php).

[3] Europa Aircraft (2004) Ltd. (2011). "Europa XS Monowheel Overview" (http://www.europa-aircraft.co.uk/monowheel/). . Retrieved 3 September 2011.

[4] The Office of the NASA Aviation Safety Reporting System (January 2004). "Gear Up Checkup" (http://asrs.arc.nasa.gov/docs/cb/ cb_292.pdf). *Call Back ASRS* (NASA) (292). . Retrieved 1 April 2012.

[5] Scislowska, Monika (3 November 2011). "Warsaw airport back to work after plane emergency" (http://www.msnbc.msn.com/id/ 45149697/ns/travel-news/#.TrLb1GCIZQE). *Associated Press via msnbc.com.* . Retrieved 13 January 2012.

[6] "Landing Gear" (http://www.biggles-software.com/software/757_tech/landing_gear/landing_gear.htm) Retrieved 10 November 2011

[7] Commander Premier Aircraft Corporation (2008). "Commander 115 Features - Major Systems" (http://home.att.net/~jBaugher1/f4_41. html). . Retrieved 2008-11-10.

[8] Hilmerby, Stellan F. (undated). "Landing Gear" (http://www.hilmerby.com/md80/md_gear.html). . Retrieved 2008-11-10.

External links

- Alphonse Pénaud website (http://www.flyingmachines.org/pend.html)
- FAA Website (http://www.faa.gov/airports/resources/publications/orders/media/Construction_5300_7.pdf)

Gnome-Rhône_Mistral_Major

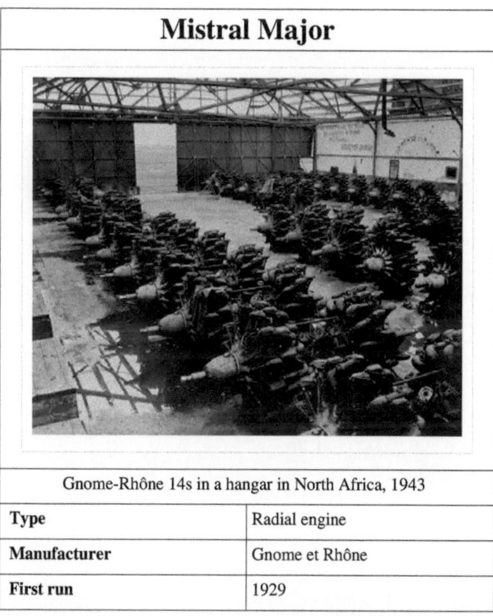

Mistral Major

Gnome-Rhône 14s in a hangar in North Africa, 1943

Type	Radial engine
Manufacturer	Gnome et Rhône
First run	1929

The **Gnome-Rhône 14K** *Mistral Major* was a 14-cylinder, two-row, air-cooled radial engine. It was Gnome-Rhône's major aircraft engine prior to World War II, and matured into a highly sought-after design that would see licensed production throughout Europe and Japan. Thousands of Mistral Major engines were produced, used on a wide variety of aircraft.

Design and development

In 1921 Gnome-Rhône purchased a license for the highly successful Bristol Jupiter engine and produced it until about 1930, alongside the smaller Bristol Titan. Starting in 1926, however, they used the basic design of the Titan to produce a family of new engines, the so-called "K series". These started with the 5K *Titan*, followed by the 7K *Titan Major* and 9K *Mistral*. By 1930, 6,000 of these engines had been delivered.

However, the aircraft industry at that time was rapidly evolving and producing much larger aircraft that demanded larger engines to power them. Gnome-Rhône responded by developing the 7K into a two-row version that became the 14K *Mistral Major*. The first test examples were running in 1929.

Licence built derivatives

- Manfred Weiss WM K.14
- Piaggio P.XI
- IAR 14K
- Tumansky M-87

Applications

- Amiot 143
- Aero A.102
- Bloch MB.200
- Bloch MB.210
- Breguet 274
- Breguet 460
- Breguet 521
- Dornier Do 17K
- Farman F.222
- Loire 46 C1
- PZL P.24
- PZL P.43
- Potez 62
- Potez 651

Aircraft powered by G-R 14K derivatives

- Aero A.102
- Breda Ba.65
- Breda Ba.88
- CANT Z.1007
- CANT Z.1011
- Caproni Ca.135
- Caproni Ca.161
- Heinkel He 70
- IAR 37
- IAR 80
- Ilyushin DB-3
- MÁVAG Héja
- Reggiane Re.2000
- Saab 17
- Savoia-Marchetti SM.79
- Savoia-Marchetti SM.84
- Sukhoi Su-2
- Weiss WM-21 Sólyom

Specifications (Gnome-Rhône 14Kdrs)

Data from [1]

General characteristics

- **Type:** Fourteen-cylinder two-row air-cooled radial engine
- **Bore:** 146 mm (5.75 in)
- **Stroke:** 165 mm (6.5 in)
- **Displacement:** 38.72 l (2,363 in³)
- **Diameter:** 1,296 mm (51.02 in)
- **Dry weight:** 540 kg (1,190 lb)

Components

- **Valvetrain:** Overhead valves
- **Supercharger:** Single-speed centrifugal type supercharger
- **Fuel system:** Stromberg carburetor
- **Fuel type:** 87 octane rating gasoline
- **Cooling system:** Air-cooled
- **Reduction gear:** 2:3

Performance

- **Power output:**
  - 743 kW (996 hp) at 2,390 rpm for takeoff
  - 821 kW (1,100 hp) at 2,390 rpm at 2,600 m (8,530 ft)
- **Specific power:** 21.23 kW/l (0.47 hp/in³)
- **Compression ratio:** 5.5:1
- **Specific fuel consumption:** 328 g/(kW•h) (0.54 lb/(hp•h))
- **Oil consumption:** 20 g/(kW•h) (0.53 oz/(hp•h))
- **Power-to-weight ratio:** 1.52 kW/kg (0.92 hp/lb)

See also

Related development

- IAR 14K
- Piaggio P.XI
- Tumansky M-87

Related lists

- List of aircraft engines

References

[1] Tsygulev (1939). *Aviacionnye motory voennykh vozdushnykh sil inostrannykh gosudarstv ([[Russian language|Russian* (http://base13.
glasnet.ru/text/aviamotory/t.htm)]: Авиационные моторы военных воздушных сил иностранных государств)]. *Moscow:
Gosudarstvennoe voennoe izdatelstvo Narkomata Oborony Soyuza SSR. .*

• Danel, Raymond and Cuny, Jean. *L'aviation française de bombardement et de renseignement 1918-1940* Docavia
n°12, Editions Larivière

Radial_engine

The **radial engine** is a reciprocating type internal combustion engine configuration in which the cylinders point outward from a central crankshaft like the spokes on a wheel. Multiple rows of radial cylinders can be used for increased capacity. The radial engine was very commonly used in large aircraft engines before most large aircraft started using turbine engines.

Since the axes of the cylinders are coplanar, the connecting rods cannot be attached to the crankshaft as usual (Inline-four_engine). Instead, the pistons are connected to the crankshaft with a master-and-articulating-rod assembly. One piston, the uppermost one in the animation, has a master rod with a direct attachment to the crankshaft. The remaining pistons pin their connecting rods' attachments to rings around the edge of the master rod.

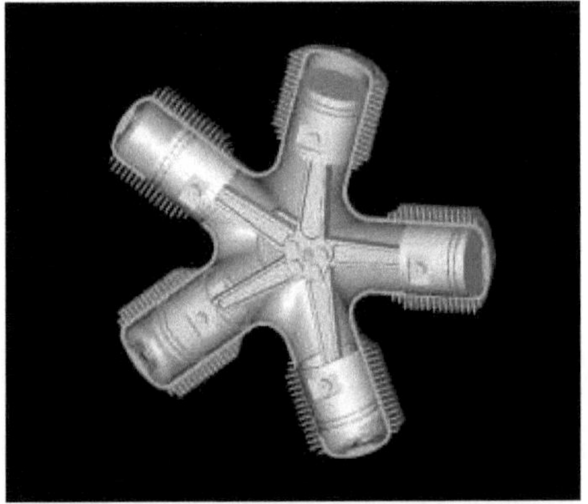

Radial engine in a cut-away view

Radial engine of a biplane

Four-stroke radials always have an odd number of cylinders per row, so that a consistent every-other-piston firing order can be maintained, providing smooth operation. For example, on a 5-cylinder engine the firing order is 1,3,5,2,4 and back to cylinder 1. Moreover, this always leaves a one-piston gap between the piston on its combustion stroke and the piston on compression. The active stroke directly helps compressing the next cylinder to fire, so making the motion more uniform. If an even number of cylinders was used, the equally timed firing cycle would not be feasible. [1] The protoype radial Zoche aero-diesels (below) have an even number of cylinders, either four or eight; but this is not problematic, because they are two-stroke engines, with twice the number of power strokes as a four-stroke engine.

The radial engine uses very few cams compared to other types. As usual for a four-stroke, the crankshaft takes two revolutions to complete the four strokes of each piston (intake, compression, combustion, exhaust). The camshaft is coaxial with the crankshaft and is typically geared with a 1:(1-n) transmission ratio, where n is the number of cylinders. As shown in the animation [2], the camshaft spins slower and in the opposite direction. The actual cams (cam lobes) are placed on two rows for the intake and exhaust. For the example, only 4 cams serve all 5 cylinders, whereas 10 would be required for a typical inline engine with the same number of cylinders.

Master rod (upright), slaves and balances from a two-row, seven-cylinder Pratt & Whitney Twin Wasp

Most radial engines use overhead poppet valves driven by pushrods and lifters on a cam plate which is concentric with the crankshaft, with a few smaller radials, like the five-cylinder Kinner B-5 and Russian Shvetsov M-11, using individual camshafts within the crankcase for each cylinder. A few engines utilize sleeve valves instead, like the very reliable 14-cylinder Bristol Hercules (built up to 1970 under licence in France by SNECMA) and the powerful 18-cylinder Bristol Centaurus.

History

C. M. Manly constructed a water-cooled five-cylinder radial engine in 1901, a conversion of one of Stephen Balzer's rotary engines, for Langley's *Aerodrome* aircraft. Manly's engine produced 52 hp (**unknown operator: u'strong' kW**) at 950 rpm.[3]

A Continental radial engine, 1944

In 1903-04 Jacob Ellehammer used his experience constructing motorcycles to build the world's first air-cooled radial engine, a three-cylinder engine which he used as the basis for a more powerful five-cylinder model in 1907. This was installed in his triplane and made a number of short free-flight hops. During 1908-9, Ellehammer developed another engine, which had six cylinders arranged in two rows of three. His engines had a very good power-to-weight ratio, but his aircraft designs suffered from his lack of understanding of control. If he had concentrated on his engines, he might have become a successful manufacturer.[4]

Another early radial engine was the three-cylinder Anzani, originally built as a "semi-radial" W3 configuration design, one of which powered Louis Blériot's Blériot XI in his July 25, 1909 crossing of the English Channel. By 1914 Anzani had developed their range, their largest radial being a 20-cylinder engine of 200 hp (**unknown operator: u'strong' kW**), with its cylinders arranged in four groups of five.[3] One of the three-cylinder "fully radial", 120° cylinder angle Anzani powerplants still exists today, in fully running condition, in the nose of Old Rhinebeck Aerodrome's restored

Pratt & Whitney R-1340 radial engine mounted in Sikorsky H-19 helicopter

and flyable 1909 vintage

Blériot XI. There is also another running Anzani at Brodhead airfield to go on a replica Blériot XI.

Radial engines are regarded as being air-cooled almost by definition—so that it is interesting that one of the most successful of the early radial engines was the Salmson 9Z series of nine-cylinder water-cooled radial engines that were produced in large numbers during the First World War. Georges Canton and Pierre Unné patented the original engine design in 1909, offering it to the Salmson company—and the engine was often known as the Canton-Unné.

42 cylinder USSR radial engine for warships

Little development of the radial engine was undertaken in Germany during World War I, where most aircraft used water-cooled inline 6-cylinder engines. Two radial engines were made there before the war but they were not proceeded with.[3]

From 1909 to 1919 the radial engine was overshadowed by its close relative, the rotary engine—which differed from the so-called "stationary" radial in that the crankcase and cylinders revolved with the propeller. Mechanically it was identical in concept to the later radial however the prop was bolted to the engine, and the crankshaft to the airframe. The primary reason for this was to ensure cooling of the cylinders, a notorious problem with all of the early radials.

In World War I, many French and other Allied aircraft flew with Gnome, Le Rhône, Clerget and Bentley rotary engines, the ultimate examples of which reached 240 hp (**unknown operator: u'strong'** kW). The German Oberursel firm (who had originated the Gnom design) made licensed copies of the Gnome and Le Rhône powerplants while Siemens-Halske built a number of their own designs including the Siemens-Halske Sh.III eleven-cylinder rotary engine.

By the end of the war the rotary engine had reached the limits of the design - particularly in regard to the amount of fuel and air that could be drawn into the cylinders during the intake stroke due to the rotary motion, while advances in both metallurgy and cylinder cooling finally allowed stationary radial engines to supersede rotary engines. In the early 1920s Le Rhône converted a number of their rotary engines into stationary radial engines although most of the other early radial engines were new designs.

By 1918, the potential advantages of air-cooled radials over the water-cooled inline engine and air-cooled rotary engine that had powered World War I aircraft were well appreciated but remained unrealized. While British designers had produced the ABC Dragonfly radial in 1917, they were unable to resolve its cooling problems, and it was not until the 1920s that the Bristol Aeroplane Company and Armstrong Siddeley produced reliable British radials such as the Bristol Jupiter and the Armstrong Siddeley Jaguar.

In the US, NACA noted in 1920 that air-cooled radials could offer an increase in the power-to-weight ratio and reliability, and by 1921 the US Navy had announced it would only order aircraft fitted with air-cooled radials while other naval air arms followed suit. Charles Lawrance's J-1 engine was developed in 1922 with Navy funding, and using aluminium cylinders with steel liners ran for an unprecedented 300 hours, at a time when 50 hours endurance was normal. At the urging of the Army and Navy the Wright Aeronautical Corporation bought Lawrance's company, and subsequent engines were built under the Wright name. The radial engines gave confidence to Navy pilots performing long-range overwater flights.[5]

Wright's 225 hp (**unknown operator: u'strong'** kW) J-5 Whirlwind radial engine of 1925 was widely acknowledged as "the first truly reliable aircraft engine".[6] Wright employed Giuseppe Mario Bellanca to design an aircraft to showcase it, and the result was the Wright-Bellanca 1, or WB-1, which was first flown in the latter part of that year. The J-5 was used on many advanced aircraft of the day, including Charles Lindbergh's Spirit of St. Louis with which he made the first solo trans-Atlantic flight.

In 1925, the American rival firm to Wright's radial engine production efforts, Pratt & Whitney, was founded. The P & W firm's initial offering, the Pratt & Whitney R-1340 Wasp, test run later that year, began the evolution of the many models of Pratt & Whitney radial engines that were to appear during the second quarter of the 20th century, among them the 14-cylinder, twin-row Pratt & Whitney R-1830 Twin Wasp, the most-produced aviation engine of any single design, with a total production quantity of nearly 175,000 engines.

In the United Kingdom the Bristol Aeroplane Company was concentrating on developing radials such as the Jupiter, Mercury and sleeve valve Hercules radials. France, Germany, Russia and Japan largely built licenced or locally improved versions of the Armstrong Siddeley, Bristol, Wright, or Pratt & Whitney radials.

Radial versus inline debate

Liquid-cooled engines often weigh more and their cooling systems are both more complex and are generally more vulnerable to battle damage. Minor shrapnel damage easily results in a loss of coolant and consequent engine seizure, while an air-cooled radial would be unaffected.[7] Additionally, radials offer higher mechanical efficiency than inline engines, as they have shorter and stiffer crankshafts, a single bank radial needing only two crankshaft bearings as opposed to the seven required for a six-cylinder inline engine of similar stiffness.[8] The shorter crankshaft also produces less vibration and hence higher reliability through reduced wear.

1935 Monaco-Trossi, a rare example of automobile use.

While a single bank radial permits all cylinders to be cooled equally, the same is not true for multi-row engines where the rear cylinders are affected by the heat coming off the front row, and air flow being masked.[9] Additionally, having the cylinders in the airflow increases drag considerably, adding turbulence that destroys the laminar airflow over the fuselage and adjacent wings. The answer to both these problems was the addition of specially designs cowlings with baffles to force the air over the cylinders. The first effective drag reducing cowling that didn't impair engine cooling was the British Townend ring or "drag ring" which formed a narrow ring around the engine covering the cylinder heads, not only reducing drag, but adding a small amount of thrust. NACA then studied the problem further and developed the NACA cowling which further reduced drag, increased thrust and improved cooling. Nearly all aircraft radial engine installations since have used NACA type cowlings. Several aircraft with liquid-cooled engines in service until the end of the Second World War, such as the Spitfire and the P-51 took advantage of the Meredith Effect to generate thrust, which worked in a similar manner to the NACA cowling.[10]

The radial engine typically has a larger frontal area, and is less amenable to streamlining and drag reduction than an inline engine. Pilot visibility is often sacrificed due to the greater width of the engine, and the designer is more limited in engine placement as greater care must be taken to ensure adequate cooling air, either in a buried engine installation or in a pusher configuration.

While inline liquid-cooled engines continued to be common until the end of World War II, radial engines saw widespread service in the successful Mitsubishi Zero and Focke-Wulf Fw 190, while the late-war Hawker Sea Fury and Grumman Bearcat, two of the fastest production piston-engined aircraft ever built, used radial engines. Until the development of the jet engine, large aircraft commonly used radial engines. Factors influencing the choice of radial over inline were reliability, simplicity in maintenance, and the ability to package a radial as a power egg, readily removable from the aircraft with the disconnection of only a few lines. Additionally, the large frontal area of these aircraft meant the radial engine's own frontal profile was a less significant factor when it came to drag.

Multi-row radials

Originally radial engines had one row of cylinders, but as engine sizes increased it became necessary to add extra rows. The first known radial-configuration engine to ever use a twin-row design was the 160 hp Gnôme "Double Lambda" rotary engine of 1912, designed as a 14-cylinder twin-row version of the firm's 80 hp Lambda single-row seven-cylinder rotary, with only the German Oberursel U.III clone of the Double Lambda reproducing the Gnome Double Lambda's twin-row design before the end of World War I. Most stationary radial engines did not exceed two rows, but the largest displacement radial engine ever built in quantity, the Pratt & Whitney R-4360 Wasp Major, with cylinders in *corncob* configuration, was a 28-cylinder 4-row

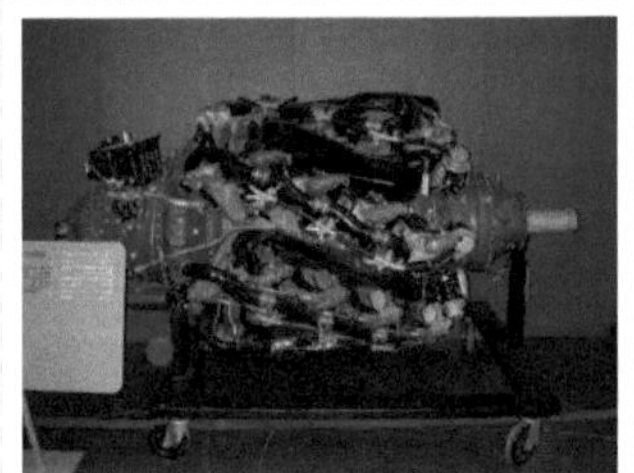

The Wasp Major, a four-row radial.

radial engine used in many large aircraft designs in the post-World War II period. The Lycoming R-7755 was the largest piston-driven aircraft engine ever produced; with 36 cylinders totaling about 7,750 in^3 (127 L) of displacement and a power output of 5000 horsepower (**unknown operator: u'strong' kW**). It was originally intended to be used in the "European bomber" that eventually emerged as the Convair B-36. Only two examples were built before the project was terminated in 1946. The USSR also built a limited number of 'Zvezda' engines with up to 56 cylinders, which were even larger in displacement than the Lycoming R-7755. The 112-cylinder diesel boat engines featuring 16 rows with 7 banks of cylinders, bore of 160 mm (6.3 in), stroke of 170 mm (6.7 in), and total displacement of 383 liters (23,931 in^3). The engine produced 10000 hp (**unknown operator: u'strong' kW**) at 2,000 rpm. They were used on fast attack craft, such as Osa class missile boats.

Modern radials

At least five companies build radials today. Vedeneyev engines produces the M-14P model, 360 hp (**unknown operator: u'strong' kW**) (up to 450 hp (**unknown operator: u'strong' kW**)) radial used on Yakovlev, Sukhoi Su-26, and Su-29 aerobatic aircraft. The M-14P has also found great favor among builders of experimental aircraft, such as the Culp's Special, and Culp's Sopwith Pup [11], Pitts S12 "Monster" and the Murphy "Moose". 110 hp (**unknown operator: u'strong' kW**) 7-cylinder and 150 hp (**unknown operator: u'strong' kW**) 9-cylinder engines are available from Australia's Rotec Engineering. HCI Aviation [12] offers the R180 5-cylinder (75 hp (**unknown operator: u'strong' kW**)) and R220 7-cylinder (110 hp (**unknown operator: u'strong' kW**)), available "ready to fly" and as a build-it-yourself kit. Verner Motor, from the Czech Republic, now builds several radial engines. Models range in power from 71 hp (**unknown operator: u'strong' kW**) to 172 hp (**unknown operator: u'strong' kW**). [13] Miniature radial engines for model airplane use are also available from Seidel in Germany, OS and Saito Seisakusho of Japan, and Technopower in the USA. The Saito firm is known for making three different sizes of 3-cylinder radials, as well as a 5-cylinder example, as the Saito firm is a specialist in making a large line of miniature four-stroke engines for model use in both methanol-burning glow plug and gasoline-fueled spark plug ignition engine formats.

Diesel radials

While most radial engines have been produced for gasoline fuels, there have been instances of diesel fueled radial engines. Two major advantages favour diesel engines - reduced fuel consumption and reduced risk of fires - however they have all the disadvantages to which diesel engines have been prone.

Packard DR-980 diesel radial aircraft engine.

Packard designed and built a diesel radial aircraft engine, the DR-980, in 1928. It was a 9-cylinder radial engine displacing 980 cubic inches and rated at 225 horsepower (**unknown operator: u'strong' kW**). On 28 May 1931, a Bellanca CH-300 fitted with a DR-980, piloted by Walter Edwin Lees and Frederick Brossy, set a record for staying aloft for 84 hours and 32 minutes without being refueled.[14] This record was not broken until 55 years later by the Rutan Voyager.[15]

The experimental Bristol Phoenix of 1928–1932 was successfully flight tested in a Westland Wapiti and set altitude records in 1934 that lasted until World War II.

A Nordberg Manufacturing Company two-stroke diesel radial engine for power generation and pump drive purposes.

In 1932 the French company Clerget developed the 14D, a 14-cylinder two-stroke diesel radial engine. After a series of improvements, in 1938 the 14F2 model produced 520 hp (**unknown operator: u'strong' kW**) at 1910 rpm cruise power, with a power-to-weight ratio near that of contemporary gasoline engines and a specific fuel consumption of roughly 80% that for an equivalent gasoline engine. During WWII the research continued, but no engines were mass-produced because of the Nazi occupation. By 1943 the engine had grown to produce over 1000 hp (**unknown operator: u'strong' kW**) with a turbocharger. After the war, the Clerget company was integrated in the SNECMA company and had plans for a 32-cylinder diesel engine of 4000 hp (**unknown operator: u'strong' kW**), but in 1947 the company abandoned piston engine development in favor of work on the emerging turbine engines.

The Nordberg Manufacturing Company of the US developed and produced a series of large two-stroke radial diesel engines from the late 1940s for electrical production, primarily at aluminium smelters and for pumping water. They differed from most radials in using having an even number of cylinders in a single bank (or row), thanks to an unusual double master connecting rod that allowed the engine to be timed so the cylinders fired in consecutive order. Variants were built that could be run on either diesel oil or gasoline or mixtures of both. A number of powerhouse installations utilising large numbers of these engines were made in the US.[16]

The Zoche[17] in Germany have produced a prototype range of radial air-cooled two-stroke diesel aircraft engines, comprising a V-twin, a single-row cross-4 and a double-row cross-8.[18] A Zoche engine has run successfully in wind tunnel tests,[19] but Zoche seem barely closer to production than they were a decade ago. Experimental engine manufacturers often experience difficulties in proceeding beyond the prototype stage due to high development and certification costs particularly with markets dominated by relatively cheap WWII era engines.

Compressed air radials

A number of radial motors operating on compressed air have been designed, mostly for use in model airplanes, or simply as a building project. Radial compressed air engines include Liney Machine Halo 5 [20], Mike Smyth SF376 [21], Taiyo UTAM4 [22], ... They have also been used in gas compressors.[23]

Use in tanks

In the years leading up to WWII, as the need for armored vehicles was realized, designers were faced with the problem of how to power the vehicles, and turned to using aircraft engines, among them radial types. The radial aircraft engines provided greater power-to-weight ratios and were more reliable than conventional inline vehicle engines available at the time. This reliance had a downside though: if the engines were mounted vertically as in the M3 Lee and M4 Sherman, their comparatively large diameter gave the tank a higher silhouette than designs using inline engines.

The Continental R-670, a 7-cylinder radial aero engine which first flew in 1931, became a widely used tank powerplant, being installed in the M1 Combat Car, M2 Light Tank, M3 Stuart, M3 Lee, LVT-2 Water Buffalo.

The Guiberson T-1020, a 9-cylinder radial diesel aero engine, was used in the M1A1E1, M2, and M3, while the Continental R975 saw service in the M4 Sherman, M7 Priest, M18 Hellcat tank destroyer, and the M44 self-propelled howitzer.

Model radial engines

A number of multi-cylinder 4-stroke model engines have been commercially available in a radial configuration, beginning with the Japanese O.S. Max firm's FR5-300 five-cylinder, 3.0 cu.in. (50 cm3) displacement "Sirius" radial in 1986. The American 'Technopower' firm had made smaller-displacement five- and seven-cylinder model radial engines as early as 1976, but the OS firm's engine was the first mass-produced radial engine design in aeromodeling history. The rival Saito Seisakusho firm in Japan has since produced a similarly sized five-cylinder radial four-stroke model engine of their own as a direct rival to the OS design, with Saito also creating a trio of three-cylinder radial engines ranging from 0.90 cu.in. (15 cm3) to 4.50 cu.in. (75 cm3) in displacement. The German Seidel firm has made both seven- and nine-cylinder "large" (starting at 70 cm3 displacement) radio control model radial engines, mostly for glow plug ignition, with an experimental fourteen-cylinder twin-row radial being tried out.

See also

- Kinner B-5
- List of aircraft engines
- Rotary engine
- Technopower
- Vedeneyev
- Warner Scarab
- Zvezda M503, a 42-cylinder Soviet missile boat diesel radial engine.

References

[1] "Firing order: Definition from" (http://www.answers.com/topic/firing-order). Answers.com. 2009-02-04. . Retrieved 2011-12-06.

[2] http://commons.wikimedia.org/wiki/File:Radial_engine_timing.gif

[3] Vivian, E. Charles (1920). *A History of Aeronautics* (http://www.daytonhistorybooks.citymax.com/page/page/3259323.htm). Dayton History Books Online. .

[4] Day, Lance; Ian McNeil (1996). *Biographical Dictionary of the History of Technology*. Taylor & Francis. p. 239. ISBN 0-415-06042-7.

[5] Bilstein, Roger E. (2008). *Flight Patterns: Trends of Aeronautical Development in the United States, 1918–1929*. University of Georgia Press. p. 26. ISBN 0-8203-3214-3.

[6] Herrmann, Dorothy (1993). *Anne Morrow Lindbergh: A Gift for Life*. Ticknor & Fields. p. 28. ISBN 0-395-56114-0.

[7] Thurston, David B. (2000). *The World's Most Significant and Magnificent Aircraft: Evolution of the Modern Airplane* (http://books.google.com/?id=7HTPRym0iYIC&pg=PA155). SAE. p. 155. ISBN 0-7680-0537-X. .

[8] Some six-cylinder inline engines used as few as 3 bearing but at the cost of heavier crankshafts, or crankshaft whipping.

[9] Fedden, A.H.R. (28 February 1929). "Air-cooled Engines in Service" (http://www.flightglobal.com/pdfarchive/view/1929/1929 - 0433.html). *Flight* **XXI** (9): 169–173. .

[10] Price 1977, p. 24.

[11] http://culpsspecialties.com

[12] http://www.hciaviation.com/

[13] http://www.vernermotor.com/

[14] Aircraft Engine Historical Society - Diesels (http://www.enginehistory.org/Diesels/CH1.pdf) Retrieved: 30 January 2009

[15] Aviation Chronology (http://www.aerofiles.com/chrono.html) Retrieved: 7 February 2009

[16] "Nordberg Diesel Engines" (http://www.oldengine.org/members/diesel/Nordberg/Nordmenu.htm). OldEngine (http://www.oldengine.org). . Retrieved 2006-11-20.

[17] "zoche aero-diesels homepage" (http://www.zoche.de/). Zoche.de. . Retrieved 2011-12-06.

[18] "Prospekt 2007.indd" (http://www.zoche.de/zoche_brochure.pdf) (PDF). . Retrieved 2011-12-06.

[19] "zoche aero-diesels testbench video" (http://www.zoche.de/Zoche_video.html). Zoche.de. . Retrieved 2011-12-06.

[20] http://www.lineymachine.com/lineyhalokit-p-2699.html?osCsid=df1711b4a444c2df252d0dd2684d6802

[21] http://home.ctlnet.com/~robotguy67/classic_cars/air_engines/V-Twin/air_engines.htm

[22] http://www.automation-dfw.com/pdf_pneu/taiyo-01utam4airmotors.pdf

[23] "Bock radial piston compressor" (http://www.bock.de/en/Product_overview.html?ArticleSizesGroupID=169). Bock.de. 2009-10-19. . Retrieved 2011-12-06.

External links

- Rotec radial engines (http://www.rotecradialengines.com/)
- Vedeneyev engines (http://www.russianaeros.com/vedenyevproduct.htm)
- Murphy Moose M-14P info (http://www.murphyair.com/Product_Info/Super/M-14.htm)
- Radial Engine Motorcycles (http://www.jrlcycles.com/page/page/4187437.htm)
- Radial Engine Automobile (http://www.deutsche-werke.de/goggo2.htm) (German)

Caproni

Industry	Aerospace
Fate	Unknown
Founded	1908
Defunct	1950
Headquarters	Italy
Products	Transport aircraft Bombers Experimental planes Air force trainers Seaplanes

Caproni was an Italian aircraft manufacturer founded in 1908 by Giovanni Battista "Gianni" Caproni. It was initially named, from 1911, **Società de Agostini e Caproni**, then **Società Caproni e Comitti**. Caproni made the first aircraft of Italian construction in 1911. The manufacturing facilities were based in Taliedo, a peripheral district of Milan.

During World War I, Caproni developed a series of successful heavy bombers, used by the Italian, French, British and US air forces. Between the world wars, Caproni evolved into a large syndicate named

Caproni Ca.316 seaplane at its moorings.

Società Italiana Caproni, Milano, which bought some smaller manufacturers. The main subdivisions were Caproni Bergamasca, Caproni Vizzola, Reggiane and engine manufacturer Isotta-Fraschini.

Between the world wars, Caproni produced mostly bombers and light transport planes. The Società Italiana Caproni ceased to exist in 1950, although one of its divisions, Caproni Vizzola endured until 1983 when it was bought by Agusta.

Aircraft

Pre-World War I

- Caproni Ca.1 of 1910 - Experimental biplane

World War I

- Caproni Ca.1 of 1914 - Heavy bomber
- Caproni Ca.2 - Heavy bomber
- Caproni Ca.3 - Heavy bomber
- Caproni Ca.4 - Heavy bomber
- Caproni Ca.5 - Heavy bomber
- Caproni Ca.18 - Observation plane
- Caproni Ca.20 - Monoplane fighter
- Caproni Ca.31 - Modified Ca.1
- Caproni Ca.32 - Modified Italian Army version of Ca.1

Inter-War Period

- Caproni Ca.30 - Postwar redesignation of Ca.1
- Caproni Ca.33 - Postwar redesignatipn of Ca.3
- Caproni Ca.34 - Postwar redesignation of proposed modified Ca.3
- Caproni Ca.35 - Postwar redesignation of proposed modified Ca.3
- Caproni Ca.36 - Postwar redesignation of modified Ca.3
- Caproni Ca.37 - Postwar redesignation of prototype ground-attack version of Ca.3
- Caproni Ca.39 - Postwar redesignation of proposed seaplane version of Ca.3
- Caproni Ca.40 - Postwar redesignation of Ca.4 prototype
- Caproni Ca.41 - Postwar redesignation of Ca.4 variant
- Caproni Ca.42 - Postwar redesignation of Ca.4 variant
- Caproni Ca.43 - Postwar redesignation of floatplane variant of Ca.4
- Caproni Ca.44 - Postwar redesignation of Ca.5 heavy bomber
- Caproni Ca.45 - Postwar redesignation of Ca.5 aircraft built for France
- Caproni Ca.46 - Postwar redesignation of Ca.5 variant
- Caproni Ca.47 - Postwar redesignation of seaplane version of Ca.5
- Caproni Ca.48 - Airliner version of Ca.4
- Caproni Ca.49 - Proposed seaplane airliner of 1919
- Caproni Ca.50 - Air ambulance version of Ca.44
- Caproni Ca.51 - Postwar redesignation of prototype of enlarged Ca.4
- Caproni Ca.52 - Postwar redesignation for Ca.4 aircraft built for Royal Naval Air Service
- Caproni Ca.56 - Airliner version of Ca.1
- Caproni Ca.57 - Airliner version of Ca.44
- Caproni Ca.58 - Postwar redesignation for re-engined Ca.4s
- Caproni Ca.59 - Postwar redesignation for exported Ca.58s
- Caproni Ca.60 *Noviplano* - Flying boat airliner prototype
- Caproni Ca.70 - Prototype night fighter of 1925
- Caproni Ca.71 - Ca.70 variant of 1927
- Caproni Ca.73 - Airliner and light bomber
- Caproni Ca.74 - Re-engined Ca.73 light bomber

- Caproni Ca.80 - Later redesignation of Ca.74
- Caproni Ca.82 - Redesignation of Ca.73*ter* variant
- Caproni Ca.88 - Redesignation of Ca.73*quarter* variant
- Caproni Ca.89 - Redesignation of Ca.73*quarterG* variant
- Caproni Ca.97 - Civil utility aircraft
- Caproni Ca.100 - Trainer
- Caproni Ca.101 - Airliner, transport, and bomber
- Caproni Ca.102 - Re-engined Ca.101
- Caproni Ca.111 - Reconnaissance aircraft and light bomber
- Caproni Ca.113 - Advanced trainer
- Caproni Ca.114 - Biplane fighter
- Caproni Ca.122 - Prototype bomber and transport
- Caproni Ca.123 - Proposed airliner version of Ca.122
- Caproni Ca.124 - Reconnaissance and bomber floatplane
- Caproni Ca.132 - Prototype bomber and airliner
- Caproni Ca.134 - Reconnaissance biplane
- Caproni Ca.161 - High-altitude experimental aircraft
- Caproni Ca.165 - Prototype fighter of 1938
- Caproni Ca.301 - Prototype fighter
- Caproni A.P.1 - Attack aircraft derivative of Ca.301
- Caproni Ca.305 - First production version of A.P.1
- Caproni Ca.306 - Airliner prototype (1935)
- Caproni Ca.307 - Second production version of A.P.1
- Caproni Ca.308 - Export version of A.P.1 for Peru
- Caproni Ca. 308 *Borea* - Airliner
- Caproni Ca.405 -- Caproni-built version of Piaggio P.32 medium bomber
- Caproni CH.1 - Prototype fighter of 1935
- Caproni Vizzola F.5 - Fighter of 1939
- Stipa-Caproni - Experimental ducted-fan powered prototype of 1932

World War II

- Caproni Ca.133 - Transport and bomber
- Caproni Ca.135 - Medium bomber
- Caproni Ca.148 - Civil-military transport version of Ca.133
- Caproni Ca.164 - Trainer and liaision and reconnaissance aircraft
- Caproni Ca.309 *Ghibli* - Reconnaissance, ground-attack, and transport aircraft
- Caproni Ca.310 *Libeccio* - Reconnaissance aircraft and light bomber
- Caproni Ca.311 - Light bomber and reconnaissance aircraft
- Caproni Ca.312 - Re-engined version of Ca.310 sold to Norway
- Caproni Ca.313 - Reconnaissance bomber, trainer, and transport
- Caproni Ca.314 - Ground attack aircraft and torpedo bomber
- Caproni Ca.316 - Seaplane
- Caproni Ca.331 - Prototype tactical reconnaissance aircraft/light bomber (Ca.331 O.A./Ca.331A) of 1940 and prototype night fighter (Ca.331 C.N./Ca.331B) of 1942
- Caproni Campini N.1 - Experimental motorjet powered aircraft of 1940
- Caproni Vizzola F.4 - Fighter prototype of 1940 with German-made engine
- Caproni Vizzola F.5bis - Proposed version of F.4 with Italian-made engine

- Caproni Vizzola F.6 -- Fighter prototype of 1941 (F.6M) and 1943 (F.6Z)

Post-World War II

- Caproni Vizzola Calif - Family of sailplanes (A-10, A-12, A-14, A-15, A-20, A-21)
- Caproni Vizzola C22 Ventura - Light jet trainer

Article Sources and Contributors

Caproni_Ca.122 *Source*: http://en.wikipedia.org/w/index.php?title=Caproni_Ca.122 *Contributors*: Magus732, Mdnavman, Rlandmann

Bomber *Source*: http://en.wikipedia.org/w/index.php?title=Bomber *Contributors*: -js-, .:Ajvol:., 213.84.20.xxx, Academic Challenger, Ahoerstemeier, Aitias, Ajraddatz, Aldis90, Alex.muller, Anclation, Andres, Andyjsmith, Anthony Appleyard, Archiespatel1998, ArgentLA, Arpingstone, Ash sul, Aspie1, Avin, Bambuway, Barek, Bbatsell, Bcorr, Bidabadi, BilCat, Blaylockjam10, Bobblewik, Booksworm, Borgx, Brian Crawford, Buster24, Cancun771, CeeWhy2, Chamal N, Chrislk02, Christinam, Conversion script, D. Wu, DARTH SIDIOUS 2, David Newton, DeLarge, Deerhitmycar, Delirium, DexDor, Discospinster, Dondon0, Drj, ESkog, Edward321, El C, Eldarone, Ellywa, Emijrp, Emoscopes, Ericd, Eug.galeotti, Excirial, Exxolon, Fl295, Fourthords, Gdr, Genghis Khan 79, Gilliam, Gjensendk, Glatisant, Glenn, Gnaeus, Gooberliberation, GraemeLeggett, Graham87, Grendelkhan, Greyengine5, Griff91, Gsl, Guardian, Halibutt, Hellomaster5, Hephaestos, Hohum, Hutcher, IVANKO, Ian Dunster, Iste Praetor, Ivan Hong, J Di, Jellyfish dave, JohnRDaily, Knotnic, Kubanczyk, L Kensington, Lexdom, Llakais, Lophoole, MLWatts, Marcus Qwertyus, Markonen, MarnetteD, Marshall Stax, Martynas Patasius, Matthew Yeager, Maury Markowitz, Mav, Meegs, Mephistophelian, MieIKIM, Mmx1, Mo7amedsalim, Monsterman333222, Moriori, Mumbo-jumbophobe, N328KF, NEMT, Neutrino 1, Nfutvol, Noahnoahj, Noodles Paine, Nooguh, Nuno Tavares, Ospalh, Oxymoron83, Pearle, Perey, Philip Baird Shearer, Pibwl, Pip2andahalf, Possum, Prestonmag, Randall uob, Randomman07, Ratsbew, RealGrouchy, Reddi, RickK, Riddley, Rlandmann, Rmhermen, Robert Merkel, Rocastelo, Roda, Rrburke, Ryanrules281, SCDBob, Sam Hocevar, Sam jervis, Samw, Sardanaphalus, Sc147, Sharkface217, ShaunMacPherson, Shell Kinney, Shizhao, Signaleer, Slightsmile, Some jerk on the Internet, SpikeToronto, Spongefrog, Srikarkashyap, Stan Shebs, Stargoat, StaticGull, Stephen G. Brown, Stewartadcock, Stewartfip, Tannin, Template namespace initialisation script, Texasfirebrand, The High Fin Sperm Whale, The Mark of the Beast, The Savage 777, Thingg, Thumperward, Timmy Appo, Tommy2010, Topher385, Tyson temby, Vegaswikian, Versus22, Vgy7ujm, Wavey, Wik, Wiki alf, Will Beback, Willking1979, Wmilner, Xanzzibar, XavierGreen, Xchgall, Ybk33, Yintan, ZapThunderstrike, Zoe, Zzuuzz, 219 anonymous edits

Prototype *Source*: http://en.wikipedia.org/w/index.php?title=Prototype *Contributors*: ABF, Adamkumpf, Alansohn, Alexius08, Allens, Andy Dingley, Angr, Anwar saadat, Auntof6, BRUTE, Bender235, Blowdart, Bobo192, C.Fred, Canderson7, Caraguacetuba, Cgruda, Christian75, Chuq, Cimnine, Closedmouth, Coffeeflower, ColinFine, Correogsk, Cubs197, Cymru.lass, D6, DRosenbach, Darkwind, DavidCary, Dboy3587, Ddmunro, Deltabeignet, Diagraph01, Diceman, Dino246, DocWatson42, Dsajga, EagleFan, Evil saltine, Evrik, Ewlyahoocom, Excirial, FRED, FatalError, Favonian, Florian Blaschke, Fluffernutter, Fnlayson, Foobaz, Fredrik, Gbear2323, Gilliam, Glenn, Greglocock, Gregpalmerx, Groucho NL, Hanacy, Head, Hektor, Hellknowz, Henrik, Heron, Hongooi, Hraefen, Humbefa, Iwinstont, JForget, JLaTondre, JaGa, Jackienaylor, Jacobko, Jesus Escaped, JetBlast, JhanCRUSH, Jim1138, JimDunning, Jj137, Jojhutton, JuJube, Juliancolton, KP-Adhikari, Kaiblankenhorn, Kaligelos, Karanne, Kingpin13, Kku, KnightRider, Komischn, Kubigula, Kwiki, Lamdk, LedgendGamer, Leonard G., Liftarn, Lightmouse, Lir, LittleDan, Lorem Ip, Lotje, Mac, Macedonian, Madhero88, MakeChooChooGoNow, Marines6241, Martarius, MattGiuca, Matthead, Mav, Mddkpp, Melsaran, Mglg, Mikael Häggström, Mike Rosoft, Milnivlek, Mistercupcake, MrOllie, Mrfausty, Muchness, Mukerjee, MyMii, Neurolysis, Nihiltres, Nik24, Nikolas Stephan, Nono64, Ohnoitsjamie, Omnipaedista, Optimist on the run, Orpherebus, PMDrive1061, Palfrey, Pearle, Pit, PureTalent101, Qwerty1234, RHaworth, RJFJR, Radagast83, RadioFan, Rafaelthethird, Rbj, Reedy, Relpquiff, Rememberway, Ripperfrvr, Rixs, Rlsheehan, Roadrunner, Rp, Ryryrules100, S2cinc, Selket, Serenity id, Shirtwaist, Siggimoo, Sir Ray, Smelialichu, Sobolewski, Spitfire, Super-Magician, Superborsuk, Syrthiss, TEG24601, Tallus, Th1rt3en, Tharanian, The Earwig, The PIPE, The Thing That Should Not Be, The singapore ministry of education sucks, TheBendster, Theroyalweman, Thierryyyyyyy, Thumperward, Tide rolls, Tintamarre, Tom harrison, Tony Fox, Trojancowboy, Tsemii, TutterMouse, Vaganyik, Vald, Vamsikrish, Veinor, VetteDude, Wazwsx, Wiglaf, Wikilawd, Wlievens, Wunderbread11, Zabdiel, Zakiwarfel, نیانم, 290 anonymous edits

Military_transport_aircraft *Source*: http://en.wikipedia.org/w/index.php?title=Military_transport_aircraft *Contributors*: Aldis90, BD2412, Bambuway, BilCat, Buffs, CommonsDelinker, CvetanPetrov1940, DITWIN GRIM, DexDor, Gene93k, GraemeLeggett, Heaven's Army, HighSpeed-X, Htroberts, Incidious, Isaacada1, Jumentodonordeste, Jury978, Karl Dickman, Kencf0618, Khalidshou, Macgroover, Marc Lacoste, Marcus Qwertyus, Michaelmas1957, MilborneOne, Mkpumphrey, Mrg3105, Mrsaad31, NiD.29, Nick-D, Ogrom, Olegvdv68, Piano non troppo, Prunesqualer, R'n'B, Reedmalloy, Rgvis, SCARECROW, Sardanaphalus, ShakataGaNai, Snlf1, The Bushranger, The PIPE, Trekphiler, Vgy7ujm, Vipul2k3, WikHead, , 56 anonymous edits

Monoplane *Source*: http://en.wikipedia.org/w/index.php?title=Monoplane *Contributors*: 7&6=thirteen, Abmac, Ahunt, Airwolf, Alma Pater, Andrew Iverson, Anittas, Arpingstone, Arrivisto, Autodidactyl, Ballon of pi, Benstown, Berserkerus, BilCat, Breez, Chochopk, Chris the speller, CommonsDelinker, Coosbane, DanMS, Darkwind, DigitalGhost, DrKiernan, Dtom, Eloy, Empowered, Enlil Ninlil, FiggyBee, Glacialfox, GliderMaven, Hohum, J.delanoy, Japanese Searobin, Joergen, Koplimek, Kubanczyk, La goutte de pluie, Lkinkade, Ludovic89, MER-C, Mactaveous, Nibios, P199, Per Honor et Gloria, Pibwl, Plasticbadge, Rcingham, Rlandmann, RottweilerCS, Salsa Shark, Sharkface217, Soundofmusicals, Styrofoam1994, Subversive.sound, Sumsum2010, The High Fin Sperm Whale, The PIPE, Tigga en, Tirkfl, TraceyR, XJamRastafire, Yekrats, 44 anonymous edits

Undercarriage *Source*: http://en.wikipedia.org/w/index.php?title=Undercarriage *Contributors*: 7poi, Agateller, Ahunt, AndreasB, ArmoredPersonel, Arpingstone, Astatine211, AussieLegend, Bart133, Bender235, BesigedB, Bonchygeez, Brendan Moody, Bryan Derksen, CIreland, CambridgeBayWeather, Capricorn42, Captain Zyrain, Chris the speller, CommonsDelinker, ConradPino, Conscious, D3, DJ Clayworth, DMacks, Daniel Quinlan, DanielHolth, Darkest tree, DexDor, Dmottl, Dolphin51, Dycedarg, E2eamon, Echuck215, Elipongo, Emerson7, Ericg, Eurosong, Exok, Falcanary, Ffishwits, FirstPrinciples, Fjournoud, Franco3450, Fvw, Gooberliberation, Graeme374, GraemeLeggett, Gunter, Hellbus, Highonhendrix, Hmains, Hooperbloob, Iceberg3k, Iediteverything, Incompetence, Ingolfson, Johntex, Kotukunui, Krellis, Lapinmies, LeadSongDog, Lowellian, Mark.murphy, Mattbrundage, Maury Markowitz, McSly, MiShogun, Minna Sora no Shita, MoRsE, Moriori, Mortenoesterlundjoergensen, MykReeve, NameIsRon, Nevilley, Niagara, NikoSilver, Nouse4aname, Omexis, P.sachin.nayak, PSE02, Pahazzard, Pibwl, Pigsonthewing, Raymondwinn, Reswobslc, Rich257, Rjwilmsi, Rlandmann, Rp:cs, SCDBob, Salam32, Scott Wilson, ShorD4000, Sp33dyphil, Stanthejeep, Stevenghetto257, Tagishsimon, Tassedethe, The PIPE, Thenoflyzone, Thisisbossi, Thomas Willerich, Thunberg, Trashbag, Unixxx, Vslashg, Wik, Wittlessgenstein, XLerate, Youngjim, 115 anonymous edits

Gnome-Rhône_Mistral_Major *Source*: http://en.wikipedia.org/w/index.php?title=Gnome-Rh%C3%B4ne_Mistral_Major *Contributors*: AMCKen, Ain92, Andrewa, BilCat, Cobatfor, Denniss, Emt147, Gampe, GraemeLeggett, Idsnowdog, Maury Markowitz, Nigel Ish, Nimbus227, Petebutt, PpPachy, Rgvis, Rlandmann, TSRL, Trekphiler, Wxtype, 36 anonymous edits

Radial_engine *Source*: http://en.wikipedia.org/w/index.php?title=Radial_engine *Contributors*: AGToth, AMCKen, Aeons, Alstrupjohn, Andrew Nutter, Andrewa, Ankon Ray, Arjayay, Arnero, Arrivisto, Atlant, Axeman89, Bagheera, Balloonguy, Bernzoil, Bigjimr, BilCat, Blandoon, Brianhe, Bryan Derksen, Chris the speller, Colin Douglas Howell, Daleh, Davygrvy, Dhollm, Dori, Dpwkbw, Duk, Ed g2s, Emt147, Encephalon, Enzedrail, Ericd, Ericg, Euchiasmus, F.bendik, Fantastic4boy, FrummerThanThou, GCarty, Geni, Glenn, Gogtjop, Gooberliberation, GraemeLeggett, Gregorio Damian Gomez, Greyengine5, Hellbus, Hooperbloob, Howcheng, Humphrey20020, Huw Powell, Iancarine, Jmc41, Joevsimp, Joffeloff, Jóna Þórunn, Kaisershatner, Kendrick7, Kieff, Kingernie, Knotnic, Kubanczyk, Lahiru k, Leandrod, Lexington50, Liftarn, Lightmouse, LilHelpa, Lizardo tx, Lockesdonkey, Longhair, Man91816, Mark.murphy, MartinezMD, Maury Markowitz, Mccraw274, Mhrogers, Michael Daly, Michael Frind, MisfitToys, Moletrouser, Morven, Moshe Constantine Hassan Al-Silverburg, Mulad, Neodarkshadow, NiD.29, Nimbus227, Nuance 4, Ospalh, Oxhop, Paul Richter, Pavithran, Petri Krohn, Pibwl, Pol098, Polpo, Psb777, Richard Arthur Norton (1958-), Rock4arolla, Rogerd, RottweilerCS, Sabedon, Salmanazar, Sardanaphalus, Segv11, Serasuna, Sfoskett, Shreditor, Siqbal, Softeis, Soundofmusicals, Stoianovici, TSRL, Template namespace initialisation script, The PIPE, TimVickers, Trekphiler, Trieste, Typ932, Vykk, Werdan7, Wizzy, Wtmitchell, Xiner, Yonatan, Youngjim, Ævar Arnfjörð Bjarmason, 164 anonymous edits

Caproni *Source*: http://en.wikipedia.org/w/index.php?title=Caproni *Contributors*: Aecis, Ahoerstemeier, Attilios, AvicAWB, Bachcell, BilCat, Carabinieri, Cmdrjameson, Ericg, FranksValli, Greyengine5, Grumpy444grumpy, LGF1992UK, Manxruler, Mdnavman, Moongateclimber, Narvalo, NorthnBound, Paul Richter, Pearle, Pibwl, Remuel, Rlandmann, Romaniantruths, Sabuell, Sardanaphalus, Spot87, Srnec, Stahlkocher1, TSRL, Template namespace initialisation script, Typ932, 6 anonymous edits

Image Sources, Licenses and Contributors

File:Gnome-Rhone 14 Mistral Major engines 1943.jpg *Source*: http://en.wikipedia.org/w/index.php?title=File:Gnome-Rhone_14_Mistral_Major_engines_1943.jpg *License*: unknown *Contributors*: U.S. Army Signal Corps

Image:Radial engine.gif *Source*: http://en.wikipedia.org/w/index.php?title=File:Radial_engine.gif *License*: unknown *Contributors*: User:Duk

Image:Radial engine WACO QCF2.jpg *Source*: http://en.wikipedia.org/w/index.php?title=File:Radial_engine_WACO_QCF2.jpg *License*: unknown *Contributors*: User:TimVickers

File:TwinWaspConRods.jpg *Source*: http://en.wikipedia.org/w/index.php?title=File:TwinWaspConRods.jpg *License*: unknown *Contributors*: User:TSRL

Image:Rotary Piston Engine 8b03632r.jpg *Source*: http://en.wikipedia.org/w/index.php?title=File:Rotary_Piston_Engine_8b03632r.jpg *License*: unknown *Contributors*: Andy Dingley, Denniss, Marcelloo, Nimbus227, Saperaud, Sfoskett, Stahlkocher, 1 anonymous edits

Image:H19 showing engine.jpg *Source*: http://en.wikipedia.org/w/index.php?title=File:H19_showing_engine.jpg *License*: unknown *Contributors*: Alaniaris, Marcelloo, Rogerd, Stahlkocher, Threecharlie

File:USSR ship radial engine.JPG *Source*: http://en.wikipedia.org/w/index.php?title=File:USSR_ship_radial_engine.JPG *License*: unknown *Contributors*: User:Neodarkshadow

Image:Monaco-Trossi1935.jpg *Source*: http://en.wikipedia.org/w/index.php?title=File:Monaco-Trossi1935.jpg *License*: unknown *Contributors*: Andy Dingley, Herranderssvensson, Kobac, Liftarn, Sporti

Image:Pratt & Whitney R-4360 Wasp Major 1.jpg *Source*: http://en.wikipedia.org/w/index.php?title=File:Pratt_&_Whitney_R-4360_Wasp_Major_1.jpg *License*: unknown *Contributors*: User:Highflier

File:Packard DR-980 USAF.jpg *Source*: http://en.wikipedia.org/w/index.php?title=File:Packard_DR-980_USAF.jpg *License*: unknown *Contributors*: Kogo, Nimbus227, Stahlkocher

Image:IMG 0648.JPG *Source*: http://en.wikipedia.org/w/index.php?title=File:IMG_0648.JPG *License*: unknown *Contributors*: User:DanielHolth

image:Caproni Ca.316.jpg *Source*: http://en.wikipedia.org/w/index.php?title=File:Caproni_Ca.316.jpg *License*: unknown *Contributors*: EH101, Ericg

Printed by Books on Demand GmbH, Norderstedt / Germany